PCR 3

The Practical Approach Series

SERIES EDITOR

B. D. HAMES
Department of Biochemistry and Molecular Biology
University of Leeds, Leeds LS2 9JT, UK

★ indicates new and forthcoming titles

Affinity Chromatography
★ Affinity Separations
Anaerobic Microbiology
Animal Cell Culture (2nd edition)
Animal Virus Pathogenesis
Antibodies I and II
★ Antibody Engineering
★ Antisense Technology
★ Applied Microbial Physiology
Basic Cell Culture
Behavioural Neuroscience
Biochemical Toxicology
Bioenergetics
Biological Data Analysis
Biological Membranes
Biomechanics—Materials
Biomechanics—Structures and Systems
Biosensors
Carbohydrate Analysis (2nd edition)
Cell–Cell Interactions
The Cell Cycle
Cell Growth and Apoptosis
Cellular Calcium
Cellular Interactions in Development
Cellular Neurobiology
Clinical Immunology
Chromatin
★ Complement
Crystallization of Nucleic Acids and Proteins
Cytokines (2nd edition)
The Cytoskeleton
Diagnostic Molecular Pathology I and II
Directed Mutagenesis
★ DNA and Protein Sequence Analysis
DNA Cloning 1: Core Techniques (2nd edition)
DNA Cloning 2: Expression Systems (2nd edition)
DNA Cloning 3: Complex Genomes (2nd edition)
DNA Cloning 4: Mammalian Systems (2nd edition)
★ Drosophila (2nd edition)

Electron Microscopy in Biology

Electron Microscopy in Molecular Biology

Electrophysiology

Enzyme Assays

★ Epithelial Cell Culture

Essential Developmental Biology

Essential Molecular Biology I and II

Experimental Neuroanatomy

Extracellular Matrix

Flow Cytometry (2nd edition)

★ Free Radicals

Gas Chromatography

Gel Electrophoresis of Nucleic Acids (2nd edition)

Gel Electrophoresis of Proteins (2nd edition)

Gene Probes 1 and 2

Gene Targeting

Gene Transcription

★ Genome Mapping

Glycobiology

Growth Factors

Haemopoiesis

Histocompatibility Testing

HIV Volumes 1 and 2

HPLC of Macromolecules (2nd edition)

Human Cytogenetics I and II (2nd edition)

Human Genetic Disease Analysis

★ Immunochemistry 1

★ Immunochemistry 2

Immunocytochemistry

In Situ Hybridization

Iodinated Density Gradient Media

Ion Channels

Lipid Analysis

Lipid Modification of Proteins

Lipoprotein Analysis

Liposomes

Mammalian Cell Biotechnology

Medical Bacteriology

Medical Mycology

Medical Parasitology

Medical Virology

★ MHC Volume 1

★ MHC Volume 2

Microcomputers in Biology

Molecular Genetic Analysis of Populations

★ Molecular Genetics of Yeast (2nd edition)

Molecular Imaging in Neuroscience

Molecular Neurobiology

Molecular Plant Pathology I and II

Molecular Virology

Monitoring Neuronal Activity

Mutagenicity Testing

Mutation Detection

★ Neural Cell Culture

Neural Transplantation

★ Neurochemistry (2nd edition)

Neuronal Cell Lines

NMR of Biological Macromolecules

Non-isotopic Methods in Molecular Biology
Nucleic Acid Hybridization
Nucleic Acid and Protein Sequence Analysis
Oligonucleotides and Analogues
Oligonucleotide Synthesis
PCR 1
PCR 2
★ PCR 3
Peptide Antigens
Photosynthesis: Energy Transduction
Plant Cell Biology
Plant Cell Culture (2nd edition)
Plant Molecular Biology
Plasmids (2nd edition)
★ Platelets
Pollination Ecology
Postimplantation Mammalian Embryos
Preparative Centrifugation
Prostaglandins and Related Substances
Protein Blotting
Protein Engineering
★ Protein Function (2nd edition)
Protein Phosphorylation
Protein Purification Applications
Protein Purification Methods
Protein Sequencing
★ Protein Structure (2nd edition)
★ Protein Structure Prediction
Protein Targeting
Proteolytic Enzymes
Pulsed Field Gel Electrophoresis
Radioisotopes in Biology
Receptor Biochemistry
Receptor–Ligand Interactions
RNA Processing I and II
★ Subcellular Fractionation
★ Signalling by Inositides
Signal Transduction
Solid Phase Peptide Synthesis
Transcription Factors
Transcription and Translation
Tumour Immunobiology
Virology
Yeast

Preface

Cellular localization of DNA and RNA sequences in cells and tissues by *in situ* hybridization (ISH) has greatly facilitated our understanding of disease processes. Conventional ISH is, however, of limited sensitivity and is also affected by the particular target being probed. Although the polymerase chain reaction (PCR) has revolutionized genetic analysis of disease by amplifying rare nucleic acid sequences easily and reproducibly from animal and plant cells, it does not allow morphological correlation. In 1990, Haase and colleagues described a new hybrid technique that coupled PCR with ISH, thereby combining morphological localization with high sensitivity detection. The technique instantly became known as *in situ* PCR, but with time has been refined into at least four different techniques: DNA *in situ* PCR and DNA PCR ISH for the detection of DNA targets in cells, and RT *in situ* PCR and RT-PCR ISH for the detection of RNA targets. A similar technique called PRINS (primed *in situ* synthesis) has also been described and is applicable to chromosomal analysis in interphase nuclei and metaphase chromosome spreads.

These techniques have enormous potential for biologists, pathologists, geneticists, etc. in allowing ready access to any DNA or RNA target within the cell and offer the investigator the ability to perform sub-genomic *in situ* analyses. Since the initial descriptions of PCR-ISH and its variants, there have been problems of reproducibility and reliability of these techniques. This has been in part due to the almost infinite variation in techniques used by individual workers and therefore the main purpose of this text is to present well-tested protocols for in-cell amplification of DNA and RNA in cells and tissues written by leading exponents of the methodology. From their accumulated knowledge and experience, we hope you will gain an insight into the field of in-cell amplification and recognize its potential in your particular area of investigation.

We thank all of the contributors and the staff at Oxford University Press for giving their time and effort in producing this book.

Liverpool and Oxford C. S. H.
January 1997 J. J. O'L.

Contents

Contributors

MUHAMMED AMJAD
Department of Medicine, Division of Infectious Diseases, Thomas Jefferson University, Philadelphia, PA 19107, USA.

OMAR BAGASRA
Centre for Human Retrovirology and Molecular Therapeutics, Thomas Jefferson University, Philadelphia, PA 19107, USA.

LISA E. BOBROSKI
Department of Medicine, Division of Infectious Diseases, Thomas Jefferson University, Philadelphia, PA 19107, USA.

M. JIM EMBLETON
Paterson Institute for Cancer Research, Christie Hospital NHS Trust, Wilmslow Road, Manchester M20 9BX, UK.

JOHN R. GOSDEN
MRC Human Genetics Unit, Western General Hospital, Edinburgh EH4 2XU, UK.

JOHN HANSEN
Department of Medicine, Division of Infectious Diseases, Thomas Jefferson University, Philadelphia, PA 19107, USA.

C. SIMON HERRINGTON
Department of Pathology, Duncan Building, Royal Liverpool University Hospital, Liverpool L69 3GA, UK.

DAVID HOPWOOD
Gastroenterology Laboratory, Department of Molecular and Cellular Pathology, Ninewells Hospital and Medical School, Dundee DD1 9SY, UK.

DAVID HOWELLS
Perkin–Elmer, Applied Biosystems, Kelvin Close, Birchwood Science Park North, Warrington WA3 7PB, UK.

ADRIAN IRELAND
Department of Surgery, St Vincent's Hospital, Elk Park, Dublin 4, Republic of Ireland.

JOHN J. O'LEARY
Formerly: Nuffield Department of Pathology, Level 4, Academic Block, John Radcliffe Hospital, Oxford OX3 9DU, UK.
Current address: Department of Pathology, Cornell University Medical College, New York, NY 10021, USA.

BRUCE K. PATTERSON
Department of Obstetrics/Gynecology and Medicine, Division of Infectious Diseases, Northwestern University Medical School, 333 East Superior Street, Suite 410, Chicago, IL 60611, USA.

STEVE PICTON
Perkin–Elmer, Applied Biosystems, Kelvin Close, Birchwood Science Park North, Warrington WA3 7PB, UK.

ROGER J. POMERANTZ
Department of Medicine, Division of Infectious Diseases, Thomas Jefferson University, Philadelphia, PA 19107, USA.

YOSHIRO SHIBASAKI
MRC Human Genetics Unit, Western General Hospital, Edinburgh EH4 2XU, UK.

DARRYL SHIBATA
Department of Pathology, University of Southern California School of Medicine, 1200 N. State Street 736, Los Angeles, CA 90033, USA.

SHIRLEY A. SOUTHERN
Department of Pathology, Duncan Building, Royal Liverpool University Hospital, Liverpool L69 3GA, UK.

ERNST J.M. SPEEL
Department of Molecular Cell Biology and Genetics, University of Limburg, PO Box 616, 6200 MD Maastricht, The Netherlands.

JACQUELINE A. STARLING
Hybaid Ltd, 111–113 Waldegrave Road, Teddington, Middlesex TW11 8LL, UK.

Abbreviations

AEC	3′-amino-9-ethylcarbazole
AIDS	acquired immunodeficiency syndrome
AMCA	aminomethylcoumarin
AMV	avian myeloblastosis virus
AP	alkaline phosphatase
APES	aminopropyltriethoxysilane
BCIP	5-bromo-4-chloro-3-indolyl phosphate
BSA	bovine serum albumin
CMV	cytomegalovirus
CNS	central nervous system
DAB	diaminobenzidine
DAPI	diaminophenylindole
DEPC	diethylpyrocarbonate
DMF	dimethylformamide
DMSO	dimethylsulfoxide
cDNA	complementary DNA
DNase	deoxyribonuclease
DTT	dithiothreitol
EBV	Epstein–Barr virus
EDTA	ethylenediamine tetra-acetic acid
EGFR	epidermal growth factor receptor
EM	electron microscopy
FAM	6-carboxyfluorescein
FISH	fluorescent *in situ* hybridization
FISNA	fluorescence *in situ* 5′-nuclease assay
FITC	fluorescein isothiocyanate
HBV	hepatitis B virus
HEX	hexachloro-6-carboxyfluorescein
HHV	human herpes virus
HIV	human immunodeficiency virus
HPV	human papillomavirus
HRP	horseradish peroxidase
HSV	herpes simplex virus
ISH	*in situ* hybridization
IVT	*in vitro* transcription
LGV	lymphogranuloma venereum
MMLV	Moloney murine leukaemia virus
MMTV	murine mammary tumour virus
NBF	neutral buffered formaldehyde
NBT	Nitroblue Tetrazolium

N-CAM	neural cell adhesion molecule
NFM	non-fat milk
NGS	normal goat serum
NHS	N-hydroxysuccinimide
OCT	optimal cutting temperature
PBMC	peripheral blood mononuclear cells
PBS	phosphate buffered saline
PCR	polymerase chain reaction
PDH	pyruvate dehydrogenase
PE	phycoerythrin
PISH	PCR *in situ* hybridization
POMC	pro-opiomelanocortin
PRINS	primed *in situ* DNA synthesis
PVA	polyvinyl alcohol
RNase	ribonuclease
ROX	6-carboxy-X-rhodamine
RT	reverse transcriptase
RT-PCR	reverse transcriptase PCR
SIV	simian immunodeficiency virus
SPPCR	solution phase PCR
SSC	saline sodium citrate
SSCP	single strand conformation polymorphism
STF	Streck tissue fixative
SURF	selective ultraviolet fractionation
T_m	melting temperature
TAE	Tris–acetate–EDTA buffer
TAMRA	5-methyl-6-carboxyrhodamine
TBS	Tris-buffered saline
TdT	terminal deoxynucleotidyltransferase
TET	tetramethyl-6-carboxyrhodamine
TMB	tetramethylbenzidine
TRITC	tetramethyl rhodamine isothiocyanate
YAC	yeast artificial chromosome

1

Fixation of tissues for the polymerase chain reaction

DAVID HOPWOOD

1. Introduction

Cell and tissue fixation was originally devised to produce good morphology. The aims of fixation have been defined variously but a number of features are in common (1,2):

- to prevent autolysis and bacterial attack
- to preserve cell shape and volume during subsequent procedures
- to retain as life-like a condition as possible
- to facilitate subsequent processing and staining

These aims are met in part and workable and reproducible morphological results achieved. Some substances, such as lipids, glycogen, and iron, are lost to different degrees during routine processing. Nonetheless, the micro-anatomy of the tissues is well enough preserved to allow a histopathological diagnosis to be established in most cases. The needs of hospital histopathology are the processing of large numbers of specimens received under a variety of conditions and from a multitude of anatomical sites. The system must be robust, simple, and reproducible. It is important to remember that it has been devised and has evolved to give good morphology rather than the optimum preservation of DNA and RNA.

Most commonly, after the tissue has been taken from the patient, it is placed immediately into fixative and sent to the pathology laboratory. This transit may take minutes to hours, depending on the portering system, the time of day, the day of the week, and public holidays. Frozen sections are sent immediately to the pathology laboratory for diagnosis as fresh unfixed tissue, whilst the patient is still anaesthetized, before the definitive operation is carried out. Tissue blocks are gathered in the pathology department in fixative during the daytime before being processed overnight through to paraffin wax.

Tissue is taken by the clinician either as an investigative biopsy or as a definitive/curative tissue specimen. The former may be 2–3 mm across, such as an intestinal mucosal biopsy. Needle cores of 1–1.5 mm × 10–15 mm are taken from liver, kidney, prostate and breast. Skin ellipses of 5 mm × 3 mm are taken through or to remove a lesion. Large specimens are dissected by pathologists and appropriate numbers of blocks (up to 25 × 20 × 3 mm^3) taken which are representative of the lesion(s) which is present.

A number of fixatives are in common use in the UK, mostly based on formaldehyde (1), which is generally buffered with phosphate to pH 7.2. Alternatively, formaldehyde may be diluted with isotonic saline or tap water as a 10% solution to give 4% formaldehyde. In continental Europe other fixatives are used commonly, e.g. Bouin in France. For electron microscopy of kidney and muscle biopsies, tissues are fixed in glutaraldehyde. Cytological preparations, e.g. smears of the cervix or fine needle aspirates, are usually fixed on the slide with a spray or air dried. The commercially available mixtures usually contain alcohol as their chief constituent. For reasons of speed of specimen turnaround, a small number of centres use microwave stabilization as the first step in processing.

A second fixative is then used in some centres or has been in the past. We have used phenol–formaldehyde (3) which speeded the process of fixation and produced no problems as far as immunohistochemistry was concerned. Previously, this department and others used mercuric chloride as a second fixative to improve staining with the trichrome methods. Various mercury-containing fixatives have been used in pathology and, therefore, before using archival material for resource purposes, it is probably worth enquiring about the methodology of fixation used, as mercury-containing fixatives have a deleterious effect on subsequent manipulation of DNA.

Special arrangements may be made with the clinicians for obtaining specimens. Here, the tissue may be received at the optimum just seconds after its removal from the patient. If animal material is being used, then timing can be much more closely controlled. Otherwise, the vagaries of the agonal phase (i.e. the period between the interruption of the oxygen/blood supply to immersion of the tissue in fixative; for human tissues obtained at surgery, this is dictated by the needs of the operation and varies from patient to patient depending on the ease or otherwise of the surgery) and portering will have to be contended with.

2. Reaction of fixatives with nucleic acids

It has been found that tissue fixation gives rise to problems for PCR, Southern blotting, *in situ* polymerase chain reaction (PCR), and related techniques (4,5).

The effects of a dozen or so fixatives on fresh tissues have been studied, with time control (6). The results show that fixatives can be divided into three groups. Acetone and neutral buffered formaldehyde gave the most usable DNA on extraction. A second group of fixatives were less satisfactory: these included paraformaldehyde, methacarn, Zamboni and acetic-formaldehyde. Fixatives containing mercuric chloride or picric acid were unsatisfactory (4). Other reports have found that formaldehyde (pH not specified) gave poor results. Ethanol on the other hand gave the most usable nucleic acid.

Fresh tissue has been found to be the most satisfactory for obtaining DNA that can be used for PCR. Tissues investigated include kidney, cervix, breast, and carcinoma of the prostate (7). Anoxic and autolytic tissues were sub-optimal for nucleic acid retrieval (4). This raises questions as to exactly what DNA is extracted from pathological lesions, which often contain necrotic and anoxic areas. The nature of the tissue is also relevant: in more compact and dense tissues the fixative takes longer to diffuse in and react.

The duration of fixation affects the amount of usable DNA that can be extracted from the tissues. This has been demonstrated in several studies (6,7). Some reports also suggest that formaldehyde-fixed tissue in paraffin blocks becomes less satisfactory after 5 years (6). Other reports have found good retrieval of nucleic acids after tissues have spent 40 years in paraffin blocks. There is a report that cloning of DNA can be achieved from tissues which have been dead for a very long time, but preserved under adequate conditions. The subject was a mummified Egyptian who had died in approximately 2600 BC (8).

There are various mechanisms whereby fixation may alter DNA. There are some reports that acid fixatives may produce small DNA fragments. Buffered formaldehyde does not do this (6). Formaldehyde reacts with tissues at pH 7 to form Schiff bases with free amino groups in nucleic acids and proteins, including histones. This may produce formylation or methylene bridges with other nucleic acids or proteins. Histones are relatively small structural proteins associated with DNA. They have a high proportion of lysine and arginine which helps them to bind tightly to the DNA. The positive charge of the amino acids results in aldehydes reacting readily with them.

These reactions may inhibit the PCR polymerase, causing it to 'jump off' the nucleic acid at a formylation point. It has been suggested that the frequency of these points may limit the size of the amplification product. Equally, these points may inhibit the degradation of RNA and DNA, although some are sensitive to Pronase (7).

There is some evidence that the cross-linking of RNA and DNA is slowly and irreversibly progressive with time. This is due to the phenomenon whereby the double strand 'breathes' or forms a single-stranded 'bubble'. At this point formaldehyde may react with the exposed nucleoside bases (7).

3. Fixation

3.1 Fixatives

The most commonly used fixatives are formaldehyde based (see *Protocol 1*). Glutaraldehyde/osmium tetroxide is used for electron microscopy and glutaraldehyde for eyes. Commercially available formaldehyde is a 35–40% gas by weight solution in water. This is known also as 'formalin' which, in fact, is a trade name. In aqueous solutions, formaldehyde exists mostly as its monohydrate methylene glycol in equilibrium with monomeric formaldehyde. Low molecular weight polymers are also present, which may be seen as a white deposit. There may also be a number of impurities present, including formic acid and methanol, plus various inorganic substances

Protocol 1. Formaldehyde-based fixatives

Equipment and reagents

- Concentrated formalin solution (40% formaldehyde)
- $NaH_2PO_4.H_2O$
- Na_2HPO_4
- NaOH
- NaCl

A. *Neutral buffered formalin*

1. Dissolve 6.5 g Na_2HPO_4 and 4 g $NaH_2PO_4{\cdot}H_2O$ in 500 ml of distilled water.
2. Add 100 ml 40% formaldehyde and make up to 1 litre with distilled water.
3. Check the pH and adjust if necessary to pH 7.2 using the appropriate phosphate salt.

B. *Formal saline*

1. Mix 100 ml 40% formaldehyde and 900 ml distilled water.
2. Dissolve 9 g of sodium chloride in the mixture.

C. *Paraformaldehyde*[a]

1. Mix 41.5 ml 2.26% sodium dihydrogen phosphate and 8.4 ml 2.25% sodium hydroxide and heat to 70–80°C in a covered container.
2. Add 2 g of paraformaldehyde and stir until it dissolves to give a clear solution.
3. Filter through a Whatman paper filter and cool.
4. Adjust the pH to 7.2–7.4.

[a] This fixative should be made up fresh for each use with Analar grade paraformaldehyde.

3.2 Diffusion

Tissue fixation depends on a number of factors. Important amongst these is diffusion of the fixative into the tissue. This is governed by a simple physical law first investigated in tissues by Medawar (9). He showed that the depth penetrated (d) was proportional to the square root of time (t). This may be written

$$d = K\sqrt{t}$$

The constant (K) is the coefficient of diffusibility. This is specific for each fixative. The coefficient of diffusibility at 1 h is the distance in millimetres that the fixative has diffused into the tissues. The coefficients have been measured using gels or uniform tissues such as liver. The values found in tissues are usually lower than gels, because of various cellular and tissue barriers posed by cell membranes and tissue septa. Fixed tissue itself also acts as an increasing barrier to diffusion. This is vividly seen in the fixation of liver or spleen. If whole organs or even 2 cm slices are placed in formaldehyde over the weekend, the centre is still unfixed on Monday. In general, fixatives diffuse slowly into tissues. This gives rise to zones which have been fixed to different extents. This can be seen readily on microscopy in terms of differential staining. The differences may be quantified. The practical outcome is that small or thin blocks of tissue are taken by pathologists. For electron microscopy, these are 1 mm cubes. For histopathology, these are $25 \times 20 \times 3$ mm^3 to fit the cassette of the tissue processor. This allows the fixative and the other solutions used in processing to gain thorough access quickly.

Tissues are inhomogeneous. Blood vessels throughout the tissue permit the more rapid penetration of fixatives through the tissue by allowing ready access deep inside the tissue block. Some tissues have many air spaces through which fixatives may theoretically diffuse easily. Other tissues, such as tendons, consist of dense collagen. Gradients of fixation will produce inhomogeneities of fixation within the tissues, certainly when shorter fixation times are used. There is some evidence that fixatives may penetrate more rapidly along tissue planes. There is also some physiological evidence, using an Ussing chamber, that fixatives penetrate rapidly through tissue as may be assessed by the inhibition of secretion of hydrogen ions. This is at variance with the morphological observations (10).

It has been shown that the duration of fixation is an important determinant of the reproducibility of PCR, Southern blots, *in situ* PCR and related techniques (7). Prolonged fixation allows reactions between formaldehyde—or other fixatives—and DNA and proteins. There is no evidence that fixation in formaldehyde causes the fragmentation of DNA.

3.3 Kinetics of fixation

The kinetics of tissue fixation have been largely worked out by researchers in the tanning industry who are interested in the effects of aldehydes on skin to produce leather. Equivalent work has been carried out on glutaraldehyde. The reaction is ideally one of pseudo-first-order kinetics, with an infinitely large concentration of fixative compared with the tissue to be fixed. With the problems of gradients and the advancing fixative front through the tissue there are obviously times when the kinetics of fixation will be complex (11).

Tissue fixation is achieved by cross-linking between proteins. The aldehydes are capable of forming bridges with the amino groups in proteins, especially lysine and arginine (12). The distance between the reactive aldehyde groups varies with the extent of polymerization. This allows the differently spaced amino groups in the tissues, especially the proteins, to be bridged.

3.4 Temperature

Surgical specimens are usually fixed at room temperature. For electron microscopy tissues are processed at 4°C. At lower temperatures, degenerative changes in tissues are slowed down, but so also is the diffusion of the fixative into the tissue and their subsequent reaction. At later phases in the processing cycle, temperatures may attain as much as 60°C to allow infiltration with paraffin wax: this may last up to 1 h. Higher temperatures (up to 100°C) are used during the rapid processing of urgent specimens and for the fixation of tissues thought to contain tuberculous lesions (1). For tissues processed using microwaves, the temperature nominally used may be 50–60°C (13,14).

3.5 Concentration of fixatives

Good morphology with minimal distortion can be achieved using a 4% formaldehyde solution or 3% glutaraldehyde (15). Artefacts are produced by higher or lower concentrations, although glutaraldehyde is effective down to 0.25%. The presence of buffer may cause the polymerization of formaldehyde.

3.6 Duration of fixation

Pathology departments aim to give a diagnostic service within 24 h of receiving surgical specimens, for reasons of patient well-being and cost. The duration of fixation is variable. At a minimum this may be 2 h. Reasons for prolonged fixation are discussed elsewhere in this chapter. Prolonged fixation inhibits enzyme and immunological activity although it will still produce satisfactory morphological results. Washing of fixed tissues in running water can restore much activity of some enzymes. There is evidence that if fixation has not exceeded 24 h, much of the formaldehyde may be removed from the tissues by washing. This is largely because the initial reaction, namely the formation of Schiff bases, is reversible by excess water.

Brains and spinal cord are, by custom, fixed for 3 weeks. This ensures that the tissue is firm enough to allow the cutting of thin slices (3 mm) needed for pathological observation and block taking.

3.7 Hydrogen ion concentration and buffers

Most fixation is carried out using 0.2 M phosphate-buffered 4% formaldehyde at pH 7.2. Some laboratories use formaldehyde in isotonic saline which is acidic. Satisfactory fixation occurs between pH 6 and 8. During fixation the cells become increasingly acidic due to anoxia and the alteration in cell metabolism. The pH of the buffer will also have an effect on the molecular conformation of the cellular components and will affect the rate of reaction between protein and aldehyde: at higher pH, the reaction proceeds more rapidly. Care must be taken to ensure that the fixative and buffer do not react with each other. The buffers used more commonly in fixation include phosphate, cacodylate, Tris, *s*-collidine and veronal acetate.

3.8 Microwave fixation/stabilization

It has been known for a long time that tissues and blood films may be fixed by heating. Cooking is in fact an extreme form of fixation. Until recently, the heat used was relatively uncontrolled, e.g. from a Bunsen burner. Microwave ovens can now be used to deliver heat in a uniform manner; microwaves penetrate into the tissues for about 2 cm from the surface (see *Protocol 2*). The heating effect is produced by the rapid movement and relaxation of polar molecules and groups throughout the tissue at a frequency of 2.45 Ghz resulting in rapid uniform heating throughout the tissue. In this way microwaves differ from conventional heating in a water bath which produces a temperature gradient across the tissue (13,14,16).

The most useful temperature range for mammalian tissue has been found to be 45–55°C which stabilizes the tissue but does not fix it. It may be stored at 4°C for some time in this state.

Protocol 2. Microwave calibration and fixation

Equipment and reagents

- Neon bulbs with any wires broken off
- Polystyrene sheet 2.5 cm thick
- 500 ml beaker

A. *Control of power with a water ballast using a neon bulb array*

1. Cut the polystyrene sheet to fit the bottom of the oven.
2. Draw a grid of squares with sides 2–3 cm long and push a neon bulb into each grid intersection.
3. Place the array on the floor of the oven.
4. Place a 500 ml beaker containing 400 ml of water at room temperature at the back right or left corner of the oven.[a]
5. Heat the oven at maximum power for 30–60 sec and record the positions of the bulbs that light up.
6. Allow the water to cool for 2–3 min and replace the warm water with unheated water. If no bulbs light up, the water load is too big.

Protocol 2. *Continued*

7. Repeat the heating cycle, reducing the water load sequentially by 50 ml until at least one cluster of three or four bulbs lights steadily.[b] This is the optimum water load for the oven.
8. Record the position of the bulbs.

B. *Stabilization and fixation*

1. Warm up the electronics for 2 min at full power before use.
2. Place a small vessel[c] containing the appropriate fluid[d] in the area of regular heating defined in Part A.
3. Determine the heating time to 50°C.[e]
4. Fix the tissues using this pre-determined time
5. Remove the tissues from the oven as soon as possible to reduce conduction artefacts.
6. Process the tissue at room temperature.

[a] The beaker should be placed consistently in the same corner used for calibration as it cannot be assumed that the optimum water load is the same for each corner.
[b] Neon bulbs light when the surrounding temperature is approximately 60°C.
[c] A small vessel should be used as microwaves only penetrate up to 2 cm
[d] Stabilization/fixation may be achieved by microwave irradiation in one of the following fluids:
- physiological saline: this can be used for light microscopy
- chemical fixative *ab initio*: this speeds up the reaction and diffusion of fixative into the tissue.
- chemical fixative after initial fixation for several hours: this speeds up formation of cross-links.

[e] Irradiation times should be <60 sec or conduction artefacts will occur; the temperature achieved should be 50 ± 5°C.

The use of microwaves in everyday pathological practice for tissue preparation has been reported from various centres. In Newcastle, UK, it is used in the control of rejection of transplanted hearts. Patients are biopsied first thing in the morning, the biopsy processed and the slides read to give a diagnosis by the afternoon. This allows the drug regime to be adjusted if necessary, with just one visit to hospital for the patient. Similarly microwaves have been used to speed up the processing of tissues and smears sent by post overnight in Holland, enabling a diagnosis to be made on the day the tissue is received.

4. Artefacts

A number of artefacts and their sources are of importance in histological and histochemical use of tissues. Factors that affect artefacts in tissues that may be used for DNA studies include:

- length of the agonal phase
- diffusion gradients in tissues

- duration of fixation
- fixative used
- disease processes (17)

If post-mortem tissue is being used, the relative times spent by the corpse at room temperature and under refrigeration will affect the quality of the DNA extracted. The time interval between death and the necropsy will be relevant. Tissues may be retrieved at shorter intervals, with proper permission being taken, by taking relevant needle cores—equivalent to biopsies in the living.

5. Hazards

There are a number of hazards, especially when handling fresh human tissue, that the worker should be aware of both for themselves and for the well-being of colleagues. Problems can largely be avoided by proper handling of tissues, chemicals, and microwave ovens, and fall into three main areas.

(a) Biological hazards: the specimen may contain unknown or unrevealed pathogens, for example tuberculosis, hepatitis B and C, or HIV, with the possibility of infection and contamination of instruments and apparatus.

(b) Chemical hazards: these include the pulmonary and cutaneous sensitivity of workers to aldehydes and the toxic effects of arsenic in cacodylate buffer.

(c) Physical hazards associated with microwave ovens: these include burns from hot irradiated vessels, formation of explosive mixtures from vapours in the oven, and sparking from metal and writing in lead pencil on frosted glassware.

References

1. Hopwood, D. (1996). In *Theory and practice of histological techniques* (ed. J.D. Bancroft and A. Stevens), 4th edn, p. 23. Churchill Livingstone.
2. Bullock, G.R. (1984). *J. Microscopy*, **133**, 1.
3. Hopwood, D., Slidders, W., and Yeaman, G.R. (1989). *Histochem. J.*, **21**, 228.
4. Dubeau, L., Chandler, L.A., Gralow, J.R., Nichols, P.W., and Jones, P.A. (1986). *Cancer Res.*, **46**, 2964.
5. O'Leary, J.J., and Doyle, C.T. (1992). *J. Pathol.*, **166**, 331.
6. Greer, C.E., Patterson, S.L., Kiviat, N.B., and Manos, M. (1991). *Am. J. Clin. Pathol.*, **95**, 117.
7. Karlsen, F., Kalantari, M., Chitemerere, M., Johansson, B., and Hagmar, B. (1994). *Lab. Invest.*, **71**, 604.
8. Paabo, S. (1985). *Nature*, **314**, 644.
9. Baker, J.R. (1960). Methuen, London.
10. Helander, H.F., Rehm, W.S., and Sanders, S.S. (1973). *Acta Physiol. Scand.*, **88**, 109.

11. Hopwood, D., Allen, C.R., and McCabe, M. (1970). *Histochem. J.*, **2**, 137.
12. Hopwood, D. (1972). *Histochem. J.*, **4**, 267.
13. Hopwood, D., Coghill, G., Ramsay, J., Milne, G., and Kerr, M.A. (1984). *Histochem. J.*, **16**, 1171.
14. Kok, L.P. and Boon, M.E. (1992). *Microwave cookbook for microscopists*, 3rd edn. Coulomb Press, Leiden.
15. Fox, C.H., Johnson, F.B., Whiting, J., and Roller, P.P. (1985). *J. Histochem. Cytochem.*, **33**, 845.
16. Login, G.R. and Dvorak, A.M. (1994). *The microwave tool book*. Beth Israel Hospital, Boston.
17. Pentilla, A., McDowell, E.M., and Trump, B.F. (1975). *J. Histochem. Cytochem.*, **23**, 251.

2

In situ genetic analysis with selective ultraviolet radiation fractionation (SURF)

ADRIAN IRELAND and DARRYL SHIBATA

1. Introduction

Genetic analysis of specific cells and tissues provides opportunities for better correlation of genotype with phenotype. Often the relative locations of cells are as important as or more important than their cytological features. Therefore, *in situ* genetic analysis may yield insights difficult to obtain by other techniques.

Selective ultraviolet radiation fractionation (SURF) is a simple technique for the isolation of histologically defined cells present in conventional microscopic tissue sections (1,2). Very small numbers of cells can be isolated rapidly with relatively crude equipment and then genetically analysed by the polymerase chain reaction (PCR), thereby allowing direct comparisons between phenotype and genotype with microscopic correlation and the ability to detect single base changes. The SURF technique differs from traditional *in situ* hybridization techniques in that it only selectively samples portions of tissue sections, but it has certain technical advantages which are presented below.

1.1 Principles of SURF

Since tissues are complex mixtures of different cell types, many investigators have used physical microdissection techniques to isolate specific cells before analysis (for example, see refs 3–5, and many others). Unfortunately, tissue microdissection can be tedious and requires great skill. The operator must be able to cut the desired region away from the undesired cells and the underlying microscope slide, and place it into the appropriate isolation tube. Multiple dissections require great care to prevent cross-contamination.

An alternative to the direct isolation of desired cells is the elimination of undesired cells, an approach which has certain technical advantages. First,

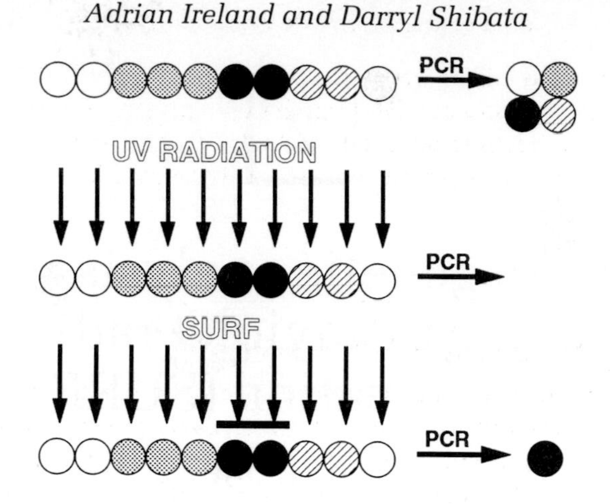

Figure 1. Principles of SURF. The PCR products of heterogeneous tissues reflect this heterogeneity (top). UV radiation prevents PCR amplification (centre). Selective UV protection leads to PCR products specifically from the protected cells (bottom).

the elimination of unwanted cells greatly reduces the chance they will contaminate the specimens of interest. Second, this approach is amenable to technical innovation. For example, a computer-controlled laser has been used to destroy everything on a microscopic slide except the small areas of interest (6). The SURF approach is less complicated and mimics the photolithotrophic techniques used for the mass production of consumer electronic microcircuits. Direct physical dissection would be an extremely inefficient way of producing small and complex integrated circuits. Instead, a photographic reproduction or mask of the desired pattern is projected on to a photosensitive material. The optical pattern is converted into a physical pattern with the exposed areas eliminated whereas the protected areas remain. Extremely fine structures can be constructed because of the high resolution of light.

Microscopic tissue sections are genetically 'photosensitive' since the DNA present in their cells can be destroyed by UV light (1,7). Therefore the cells of interest can be masked and UV light can be used to eliminate the DNA present in all other unprotected, undesired cells. The resolution of this approach can theoretically be greater than that of conventional microscopy since UV light has a shorter wavelength than visible light, hence the technique of SURF (see *Figure 1*).

1.2 Advantages of SURF

One primary advantage of SURF is that it separates the PCR from the challenge of tissue localization. Therefore, the PCR assay can be optimized with the usual techniques, and special PCR primers and conditions specific to *in situ* analysis are unnecessary. Topographical information is retained since the location of each cell fraction is documented with SURF. Questions involving novel genetic loci can be immediately applied to *in situ* analysis since conventional PCR techniques used for bulk DNA analysis are suitable with SURF. In particular, single-copy loci are easily analysed.

2. SURF

2.1 Suitable tissues and PCR targets

SURF is best performed on conventional, formalin-fixed, paraffin-embedded tissues since the DNA in such tissues is partially degraded but still sufficiently intact for PCR. PCR targets should be <200 bp. PCR with targets of 120–160 bp can usually amplify single-copy loci from >90% of formalin-fixed tissues. Shorter PCR targets allow amplification from even more degraded DNAs. However, the subsequent destruction by UV light is more difficult with shorter PCR targets. Therefore, PCR targets suitable for SURF should be >100 bp (to facilitate destruction by UV) but <200 bp (to overcome the degradation inherent to tissue fixation).

Tissues are sometimes preserved with other fixatives which inhibit subsequent PCR analysis (see also Chapter 1). These fixatives include B-5, Zenker's, Carnoy's, and Bouin's (8). Optimal fixatives are 10% neutral buffered formalin (almost universally used for human tissues) and ethanol-based fixatives. Prolonged fixation (>5 days) in formalin degrades the DNA but the usual overnight fixation typical of clinical practice yields amplification in >90% of cases. Very old paraffin blocks can be used (9).

2.2 Staining of sections

The topographic information inherent to well selected and preserved histological sections is maintained with the SURF technique. Certain modifications are necessary to facilitate genetic analysis. One primary difference is the conversion from glass to plastic microscopic slides. Glass slides have been used since the turn of the century. Plastic slides have the optical properties of glass, but the attached tissue and plastic can be cut by scissors. Specific isolation of tissue is greatly facilitated by this modification since the investigator can be confident that the desired tissues are indeed present in the proper tubes. Rapid isolation of macroscopic tissue regions is also facilitated by the plastic slides since the appropriate regions can be readily and directly identified, cut out, and placed in their respective microcentrifuge tubes.

Plastic slides are currently not commercially available. Flat 'acetate' sheets about 0.1 mm in thickness can be purchased in art and stationery stores. The production of plastic slides from these sheets is detailed in *Protocol 1* and their use for section mounting in *Protocol 2*.

Protocol 1. Manufacturing plastic microscope slides

Equipment and reagents

- Plastic 'acetate' sheets (about 0.1 mm thick)
- Paper cutter or scissors
- Slide holder
- 0.1% Poly-L-lysine adhesive solution (catalogue no. P8920, Sigma Diagnostics, St Louis, MO)

Method

1. Using sterile scissors or a dedicated paper cutter, and wearing gloves, cut the acetate sheets into pieces the same shape and size as microscope slides.
2. Place the plastic slides in the slide carrier. Slide carriers that hold slides vertically with the longer side upwards appear to work best since they support the plastic slides along three edges.
3. Dip the plastic slides in the carrier into the 0.1% poly-L-lysine solution for 5 min at room temperature, shake to eliminate most of the drops, and then air dry.
4. Store the coated slides at room temperature stacked next to each other in a sterile box.

Next, the paraffin sections are placed on the plastic slides, as described in *Protocol 2*.

Protocol 2. Sectioning of tissues on to plastic slides

Equipment and reagents

- Plastic slides (see *Protocol 1*)
- Microtome
- 'Sharpie', fine-point marking pen (Sanford Corporation, Bellwood, IL)

Method

1. Cut sections of conventional histological thickness (approximately 4 μm) and float them out in a water bath as with conventional glass slides.[a]
2. Label the slides with a 'Sharpie' marking pen.
3. Store the slides at room temperature in clean, flat, cardboard slide trays (one per well) or on edge in plastic slide cases (not touching each other).

[a] Histology technicians usually have no problem substituting plastic for glass slides. Gloves are not necessary as long as the slides are handled by the edges.

The tissue sections on the plastic slides are then stained according to *Protocol 3*.

Protocol 3. Staining of plastic slides

Equipment and reagents

- Flat metal tray
- Oven at 90°C (±5°C)
- Slide holder
- Clear-Rite 3 (Richard-Allen Medical, Richland, MI) or similar xylene substitute
- Ethanol, 100% and 95%
- 0.5% Ammonium hydroxide solution (1:200 with distilled water)
- Haematoxylin
- Eosin

Methods

1. Place the plastic slides (tissue face up) on a flat metal tray over small (1–3 mm) water drops.[a]
2. Heat the slides on the metal plate in a 90°C oven for 5–8 min. This baking melts the paraffin and makes the tissue adhere firmly to the slide. The slides should remain flat.
3. Remove the metal tray containing the slides and allow them to cool to room temperature.
4. Remove each plastic slide, place it in a slide holder, and air dry it completely.[b]
5. Stain the slides with conventional histological haematoxylin and eosin reagents as follows[c]:
   (a) Place slides in Clear Rite 3 for 3 min to deparaffinize.
   (b) Dip five times in 100% ethanol.[d]
   (c) Dip five times in 95% ethanol.
   (d) Dip four times in water.
   (e) Dip in haematoxylin three times to stain.
   (f) Dip five times in water.
   (g) Dip twice in 0.5% ammonium hydroxide to darken stain.
   (h) Dip twice in water.
   (i) Dip three times in eosin to stain.
   (j) Dip six times in 95% ethanol.
   (k) Dip six times in 100% ethanol.
   (l) Shake and air-dry for 2 min.
6. Store the stained slides at room temperature.[e]
7. Clean the metal tray by wiping with Clear-Rite and then with water. The staining solutions can be re-used until dirty. As with all pre-PCR

Protocol 3. *Continued*

procedures, the staining area should be physically isolated from PCR products.

[a] The water drops provide surface tension to prevent curling of the slides during baking.
[b] The plastic slides should be placed in alternate positions since they are less stiff than glass slides and will tend to adhere to each other during staining. Slide holders which hold slides vertically (longer side up) appear to work best since the plastic slides are supported along three sides.
[c] For tissues that tend to fall off, each step may be shortened since only light staining is necessary.
[d] In this and subsequent steps each dip is for approximately 2–4 sec.
[e] As coverslips are not used, there is some loss of histological detail. However, the histological features are usually adequate to distinguish between cells of different phenotype. Tissue folds should be avoided as these may protect the underlying tissue from UV destruction (see *Protocol 7*). However, minor imperfections are tolerated.

2.3 Dotting of selected cells

SURF requires some practice and modification of existing tools. One primary requirement is the ability to recognize histological features, so ideally the technique should be carried out by pathologists, although most individuals can achieve competence after several days or weeks of study. SURF places the emphasis on which areas to analyse since the subsequent isolations are greatly simplified.

Various UV protective ink 'umbrellas' have been tried without success as

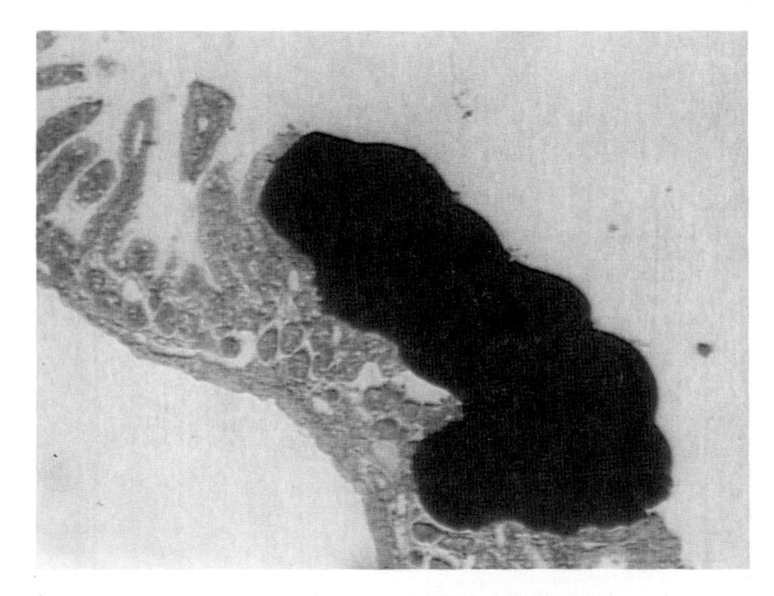

Figure 2. Ink dots placed on a strip of mucosa. The cells were protected by seven smaller dots.

many inks contain substances which inhibit subsequent PCR. The ink in 'Sharpie' (Sandford Corp, Bellwood, IL) marking pens works well as it blocks UV radiation, produces sharp dots, and does not inhibit PCR. Dots can be placed manually using a fine-point 'Sharpie' pen. However, a micromanipulator makes it much easier to place dots on the desired areas (see *Figure 2*) and is described below (see *Protocols 4–6*).

Protocol 4. Modified camera shutter

Equipment

- Camera shutter cord, 0.3–0.5 m long
- Small magnet (2 × 2 × 4 mm^3)

Method

1. Shape the camera end of the shutter cord such that a pipette tip will fit.
2. Solder the small magnet on to the tip of the shutter cord. This magnet allows a direct link between the manual movements of the shutter cord and the fishing line of the modified pipette tip (see *Protocol 5*).
3. Mount the camera shutter cord on the micromanipulator attached to the inverted microscope stage.
4. Attach the modified pipette tip (see *Protocol 5*) to the camera shutter cord allowing the fishing line to be moved directly with the shutter cord mechanism. This set-up facilitates the sterile attachment and detachment of the modified pipette tips (see *Figure 3*).

Protocol 5. Construction of modified pipette tip or dotter

Equipment

- Gel loader pipette tips (for example 25 mm capillary gel loading pipette (1–200 μl), catalogue no. 53509-015, VWR Scientific, West Chester, PA 19380
- Monofilament fishing line (one pound test)
- Stainless steel staples
- Aluminium foil
- Scissors
- Autoclave

Method

1. Shape the staples into C-shaped staples.
2. Tie the fishing line on to one end and cut to about 4 cm in length.
3. Trim the pipette tip such that its thin portion is approximately 1 cm long.
4. Thread the line with attached staple into the pipette tip.
5. Cut off any excess line such that it will be 2–4 mm inside the tip when used in *Protocol 6*.

Protocol 5. *Continued*

6. Completely cover the tip with enclosed staple and fishing line with aluminium foil. Induce a 60–80° bend between the pipette tip base and its tip and use the aluminium foil to maintain this bend.
7. Autoclave for 1 h. After cooling, the modified pipette tips will be sterile and the bend will be permanent.
8. Place the dotter on to the modified camera shutter cord (see *Protocol 4*). If necessary, both the fishing line and the outer plastic pipette sheath can be modified with sterile scissors (see *Figure 3*).

Protocol 6. Placement of protective ink dots

Equipment and reagents

- Inverted microscope with mechanical stage (for example Olympus, Model CK2 with 2×, 4×, 10×, and 20× objectives)
- Video camera (for example Sony, Model DXC-960MD 3CCD colour video camera) and recorder (or other suitable photodocumentation system)
- Micromanipulator with *x*, *y*, and *z* movements (e.g. Narishige, Model MN-3)
- Modified camera shutter cord (see *Protocol 4*) and modified pipette tip (see *Protocol 5*)
- Plastic slide with stained tissue (from *Protocol 3*)
- Glass microscope slides
- Scotch tape
- 'Sharpie', fine-point, black marking pen (Sanford Corporation, Bellwood, IL)

Method

1. Break a Sharpie pen in two manually at the grey and black junction of its plastic parts.
2. Squeeze a single drop of ink out of the internal fibre wick on to a clean glass slide. The ink is usually too dilute and must be concentrated with partial air drying for approximately 5 min. This drying time may be modified to ensure that the ink dots completely block visible and UV light and yet are not too viscous.
3. Place the modified pipette tip (see *Protocol 5*) on to the modified micromanipulator (see *Protocol 4*) and fill it by capillary action by placing its tip into the ink drop for several seconds with the micromanipulator. Only 3–5 mm of ink are necessary.
4. Place the modified pipette tip in approximately the centre of the field of view and about 2–3 mm above the tissue section using the micromanipulator.
5. Test the dotter by dotting a blank slide. Adjust the height and angle of attack of the pipette tip to ensure reproducible small dots.
6. Tape the plastic slides on to conventional glass slides, tissue side up.
7. Move the tissue regions to be dotted under the modified pipette tip

using the microscope's mechanical stage. It is generally easier to move the slide and not the dotter.

8. Place dots directly on to the small tissue regions. Usually several small dots (2–10) are placed such that approximately 100–400 cells are covered (see *Figure 2*).[a]
9. Document the exact histology protected by each dot using the video camera and recorder.[b]

[a]Placement of the dots should not contaminate the dotter since the cells and their DNA are firmly attached to the plastic slide. The monofilament fishing line is too flexible to scratch or detach the tissue easily. A single dotter can be used to protect 20–50 tissue regions. Eventually the ink will dry and clog the tip. Occasionally, the dotter will lift the tissue off the slide: this error can be observed under the microscope and the dotter should then be considered contaminated and replaced.

[b]Film-based cameras are also adequate but it is difficult to archive the many images generated by SURFing. Alternatively, topography can be documented by photocopying, ideally with a magnification of 2x (see *Figure 4*). Numbers are then used to identify each dot.

2.4 Ultraviolet irradiation

Short-wave UV radiation destroys DNA and prevents subsequent PCR. The dots placed on the tissue sections selectively protect the DNA of the cells of interest whereas the unprotected and undesired DNA is eliminated from subsequent analysis. The optimum time of UV exposure will increase as the transilluminator ages (from 90 min for a new transilluminator to 3–4 h with an older one). The UV exposure must also be increased if shorter PCR targets (<120 bp) are analysed. The protection by the dots is almost complete so excess UV exposure does not appear to be a problem. DNA can still be amplified from protected tissues accidentally exposed overnight to the UV radiation. A typical procedure is given in *Protocol 7*.

Protocol 7. Ultraviolet irradiation

Equipment

- UV transilluminator, 254 nm wavelength
- UV protection goggles

Method

1. Remove the plastic slides from their glass slides and then place them directly (tissue side down) on to the UV transilluminator.
2. Expose slides to the UV radiation for approximately 2 h.
3. Move the slides every 20–30 min to different positions on the transilluminator to ensure uniform exposure. Some plastic slides will progressively turn brown–grey with exposure, providing a visual check of the amount of UV radiation delivered. Use UV protection goggles.

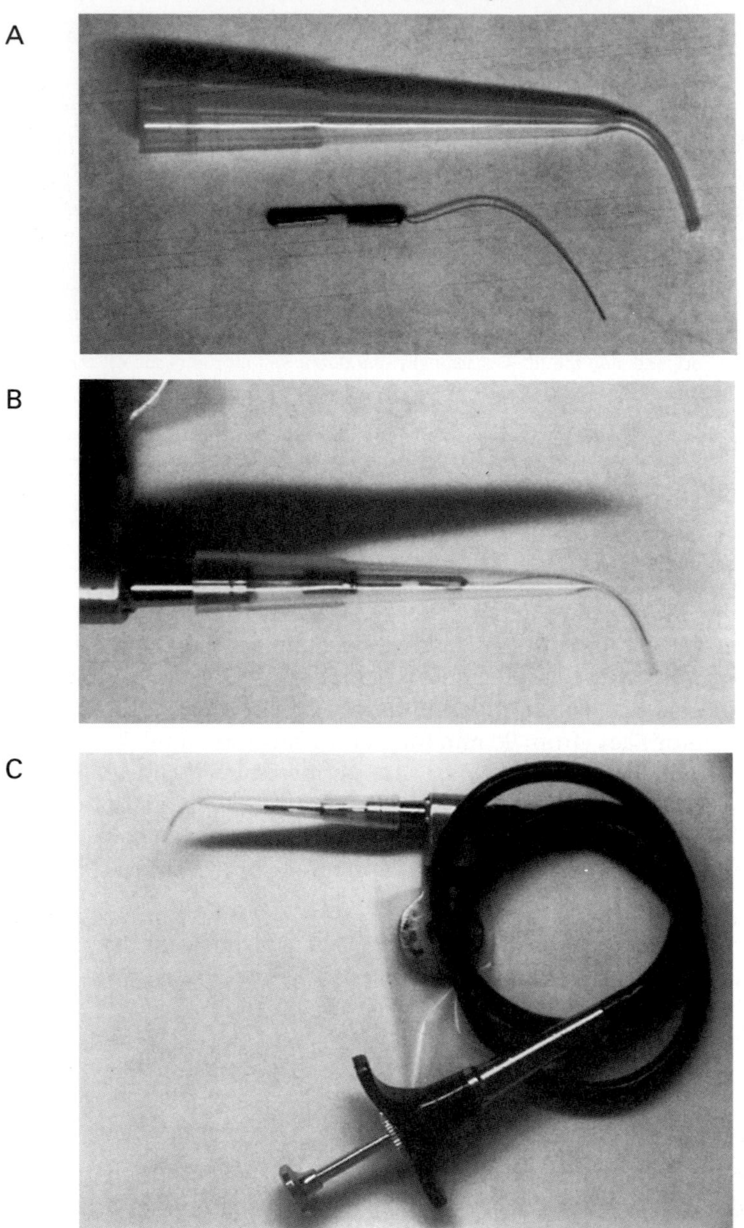

Figure 3. A simple apparatus for placing protective dots. (A) Monofilament fishing line (one pound test) is tied to a staple and placed inside a gel loader pipette tip. The bend (already present in the photograph) is induced after autoclaving. (B) The modified pipette tip is attached to the end of the camera shutter with the staple attached via a magnet. (C) The camera shutter is attached to an inverted microscope via a micromanipulator. Pressing the camera shutter directly controls the movement of the fishing line and therefore the placement of the dots.

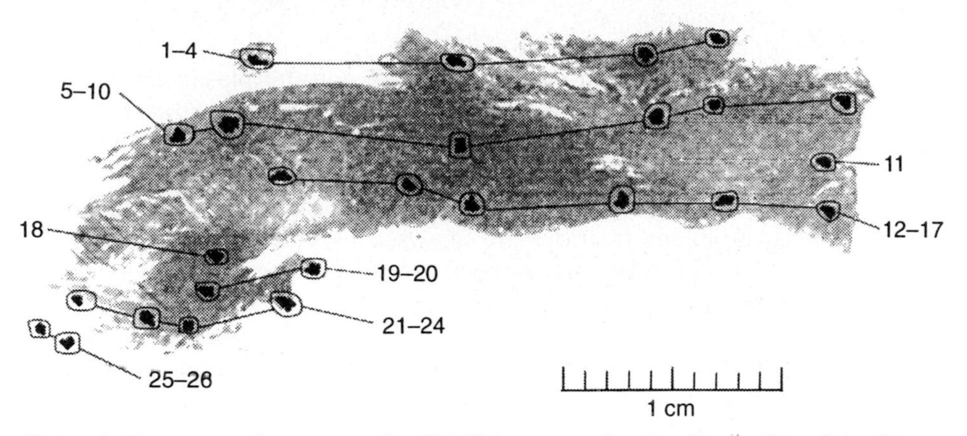

Figure 4. Photocopy of a tumour after SURF demonstrating the distribution of the dots. Each dot receives a number and subsequently a microscopic phenotype and a genotype (see *Figure 6*).

2.5 Isolation of DNA (*see Protocol 8*)

Plastic slides greatly facilitate final DNA isolation since the dots can be simply cut out with sterile scissors and placed directly into microcentrifuge tubes.

Protocol 8. DNA extraction

Equipment and reagents

- Sterile scissors
- Sterile forceps
- Sterile paper (we use photocopier paper directly from the package)
- Sterile 0.5 ml microcentrifuge tubes
- Microcentrifuge
- Water bath
- Extraction solution: add 0.7 μl of 20 mg/ml proteinase K to 33 μl of 100 mM Tris–HCl, 2 mM EDTA pH 8.0 per sample

Method

1. Cut out the plastic with attached dots (2–4 mm squares) with sterile scissors and place them directly with forceps into individual 0.5 ml microcentrifuge tubes. Work over sterile paper.
2. Add 33 μl of extraction solution to each tube.
3. Pulse them in a microcentrifuge at 13 000*g* for a few seconds.
4. Incubate overnight at 42 °C or for 4 h at 56 °C.
5. Boil the tubes for 5–7 min, vortex them, and pulse them in a microcentrifuge at 13 000*g*.
6. Before PCR, scrape the ink dots off the plastic slide with a pipette tip to ensure complete extraction.
7. Store at −20 °C.

2.6 PCR

If desired, the entire fraction including the ink dot and plastic can be subjected to PCR. Typically 8–10 μl (of the 33 μl solution from *Protocol 8*) are used in a 50 μl PCR so that multiple loci can be analysed from the same dissected cells. Radioactive techniques and a large number of PCR cycles (36–48) are usually needed since the number of starting molecules is low. The PCR products from the fractions are analysed by conventional techniques including dot blot hybridization, restriction enzyme digestion, single strand conformation polymorphism (SSCP), direct sequencing, and electrophoresis (such as microsatellite size analysis) (10,11). Loss of heterozygosity studies are also possible but require careful attention to PCR conditions since many PCR cycles are necessary to detect the low numbers of starting molecules.

Virtually any target can be analysed with SURF and PCR. The biggest concern is the size of the PCR product. It must be short enough (<200 bp, and ideally <160 bp) to be preserved in the fixed tissue but long enough (>100 bp) to be readily inactivated by UV radiation.

As with any procedure, controls are extremely important. The following controls should be used:

(a) No dot control: an adjacent square of similar but unprotected tissue should also be isolated and analysed by PCR. It should demonstrate no PCR products. If PCR products are detected with this negative control, a longer PCR target or greater exposure to the UV radiation is necessary.

(b) Duplicates: each tissue section should be SURFed at least twice to verify the distribution of each mutation.

(c) Positive controls: if difficulty is encountered in getting detectable PCR products, tissue not exposed to UV radiation should be amplified to verify that its DNA is intact. A shorter PCR target usually corrects this problem unless the tissue has been fixed in B-5, Bouin's, or Zenker's.

2.7 Data presentation and analysis

SURF yields a wealth of data which can be difficult to present. For each tumour, multiple phenotypes and locations can be correlated with multiple genotypes. Typically the raw data are collected as in *Figure 4*. The final analysis can be presented in one dimension as in *Figure 5*. In this analysis, the presence or absence of a given genotype is plotted in comparison with the different histological phenotypes present in a single tumour. Heterogeneity is noted by the transition in genotype within or between different histological regions. The genotypic transitions between histological phenotypes are very important indications of when they occurred and their possible roles in tumour progression.

The data can also be presented in two dimensions by reproducing the histological image then plotting the genotype directly on to the locations analysed by SURF. This can be done by photocopying the slide and then scanning the

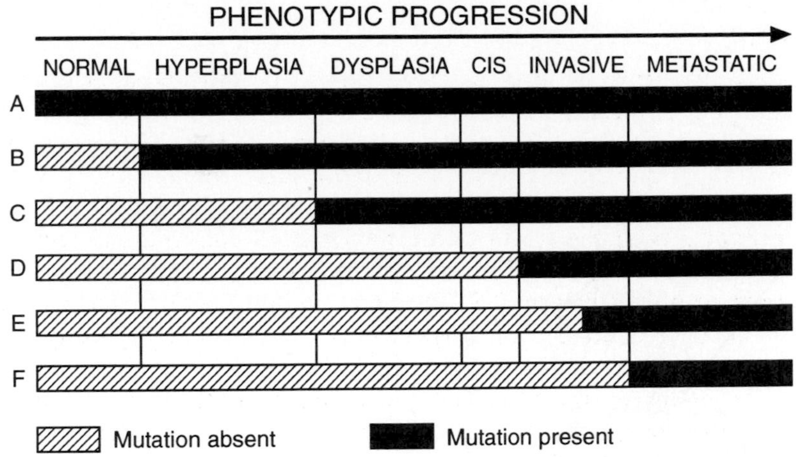

PHENOTYPIC PROGRESSION

NORMAL HYPERPLASIA DYSPLASIA CIS INVASIVE METASTATIC

A
B
C
D
E
F

Mutation absent
Mutation present

Figure 5. Possible relationships between genotype and phenotypes within a single tumour. Pattern A indicates a germline mutation whereas pattern F indicates a mutation specifically associated with metastasis. Pattern E indicates heterogeneity with the mutation probably having been acquired during tumour invasion.

image into a computer. The histological sections can be manipulated with graphics software (such as Microsoft PowerPoint) and the genotype can be placed directly on to the section. All of the data collected by SURFing can be displayed with this form of data presentation (see *Figure 6*).

3. Applications

Before PCR, genetic analysis generally required large amounts of fresh tissue since hybridization probes required weeks to detect 1–10 μg of DNA. Because of this limitation, it was impractical to analyse possible tissue heterogeneity at the resolution of single cells. Now, over a decade after the PCR revolution, large amounts of DNA are still extracted from bulk tissues even though PCR allows the genetic analysis of small numbers of molecules. It is both unnecessary and unwise to perform genetic studies on bulk extracted DNA unless it can be safely assumed that tissue heterogeneity is either not present or unimportant.

One specific example of tissue heterogeneity is multistep tumour progression. Tumours consist of mixtures of normal inflammatory and stromal cells as well as mixtures of different tumour grades. Multistep tumour progression predicts a genetic basis for the different tumour phenotypes with a greater number of oncogenic mutations in more 'advanced' tumour cells. This prediction of tumour heterogeneity can be directly tested with SURF since it is easy to determine if a specific mutation is present throughout the tumour or only in a portion (see *Figure 6*). Surprisingly, it has been difficult to detect

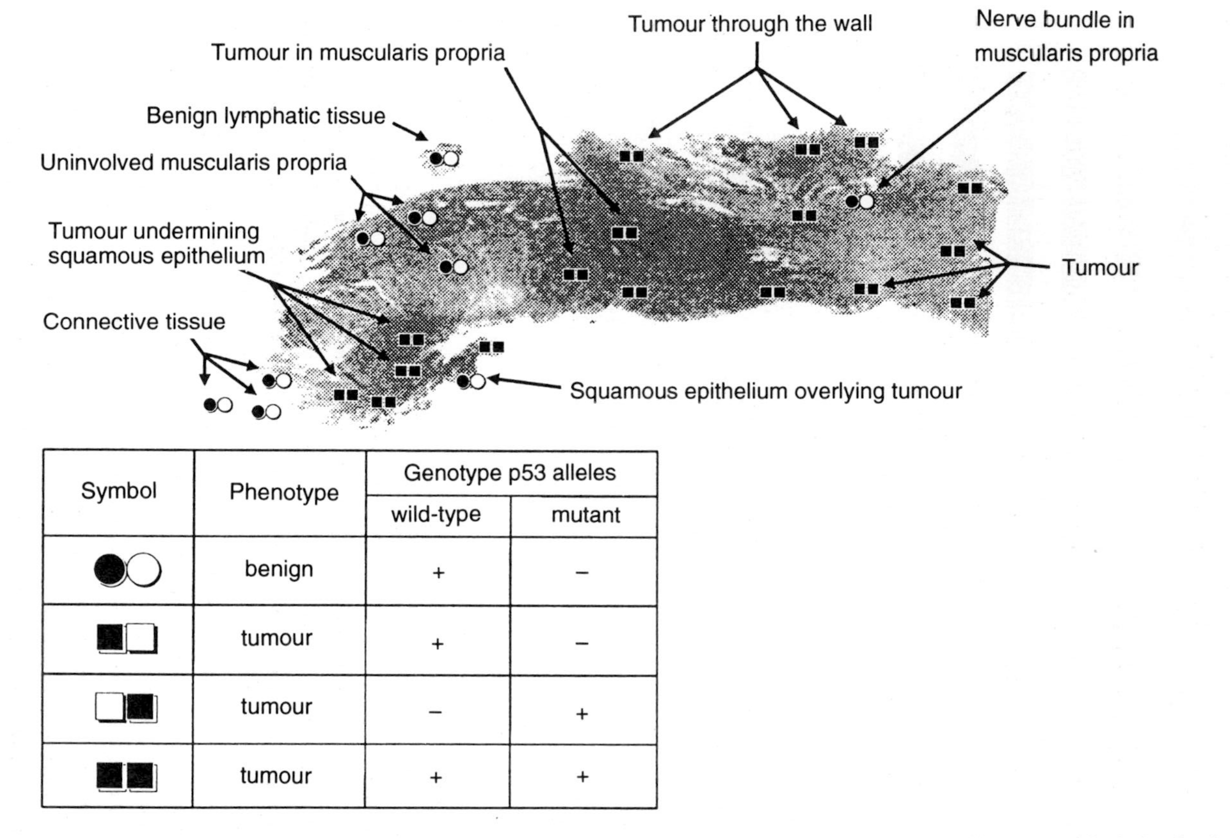

Symbol	Phenotype	Genotype p53 alleles	
		wild-type	mutant
●○	benign	+	–
■□	tumour	+	–
□■	tumour	–	+
■■	tumour	+	+

Figure 6. SURF data displayed in two dimensions. Microscopic phenotype, genotype (i.e. p53 status), and topographical distributions are shown. In this tumour, a p53 mutation without evidence of loss of the wild-type allele is present throughout the tumour, suggesting that the p53 mutation was acquired before the bulk of clonal expansion.

genetic tumour heterogeneity. In general, if a somatic mutation is present in a tumour, it will be present in all tumour regions. This distribution suggests most mutations are acquired prior to the bulk of clonal expansion.

3.1 Tumour evolution: out-of-Africa, out-of-adenoma

Although multistep tumour progression forms the theoretical basis of human cancer (12), it is very difficult to observe this process directly in humans. Besides being unethical, this approach suffers from the disadvantage that early lesions are likely to be small, and sampling would eliminate the possibility of further evolution. Therefore, the most common approach to 'observe' tumour progression is to examine multiple tumours from multiple individuals. The frequencies of a given mutation are compared between tumours of different stages. A mutation is considered to be acquired early in tumour progression if its frequency is the same at all stages. A mutation is considered to be acquired late in tumour progression if its frequency increases with stage. Although this approach is commonly employed, it has several problems. Although more mutations are generally present in later-stage tumours, a single well-defined sequence of mutations has not been defined for most tumours, which probably reflects the genetic and environmental heterogeneity of the patients included in studies.

Given that the evolution of each tumour is likely to be distinct, how can one study a process which cannot be directly observed? One general solution is to translate population genetic approaches to cancer. A tumour is not a single well-defined entity but rather a population of cells. The genotype of a tumour may be difficult to define rigorously since it is unlikely that every tumour cell is genetically identical.

Population genetics can take advantage of the expected genetic divergence of individuals over time to study historical patterns of evolution and migration. For instance, by analysing the genotypes of current and distinct human populations present in different parts of the world, it is possible to reconstruct our emergence from Africa approximately 100 000–200 000 years ago (13). Note that it would be impossible to perform this analysis without preserving the topological relations of the different populations. Tumours arise from a single initiated cell, expand, and eventually spread to different parts of the body. With techniques like SURF it is possible to consider tumour sections as continents and tumour genomes as individuals. Mutations present in different tumour parts provide clues to tumour growth. We have taken this approach to study the evolution of individual mutator phenotype tumours (14). Mutator phenotype tumours lack DNA mismatch repair and the mutation rate is greatly increased at simple repeat sequences or microsatellite. By using microsatellites as molecular tumour clocks it is possible to 'observe' past patterns of tumour evolution using a population genetic approach. In somewhat of a paradox, older tumour regions are expected to accumulate a greater number of different mutations compared with younger tumour regions ('diversity equals antiquity'), although more mutations are likely to be present within an individual cell in younger, more oncogenic regions.

Although tumour growth cannot be systematically observed in humans, the process of growth inevitably leaves a genetic legacy involving both time and space. By using *in situ* techniques of genetic analysis like SURF, it may be possible to dissect the genetic trails left in individual human tumours. Perhaps the past of a tumour will provide critical information on its possible causes and possible future behaviour.

Acknowledgements

Darryl Shibata thanks the many talented students and technicians who have gone 'SURFing' in his laboratory, including Zhi-Hua Li, Jenny Tsao, Qiuping Shu, Susan Leong, and Drs Bridgette Duggan and Antonio Hernandez.

References

1. Shibata, D., Hawes, D., Li, Z.H., Hernandez, A.M., Spruck, C.H., and Nichols, P.W. (1992). *Am. J. Pathol.*, **141**, 539.
2. Shibata, D. (1993). *Am. J. Pathol.*, **143**, 1523.
3. Meltzer, S.J., Yin, J., Huang, Y., McDaniel, T.K., Newkirk, C., Iseri, O., Volgelstein, B., and Resau, J.H. (1991). *Proc. Natl Acad. Sci. USA*, **88**, 4976.
4. Bianchi, A.B., Navone, N.M., and Conti, C.J. (1991). *Am. J. Pathol.*, **138**, 279.
5. Volgelstein, B., Fearon, E.R., Hamilton, S.R., Kern, S.E., Preisinger, A.C., Leppert, M., Nakumura, Y., White, R., Smits, A.M.M., and Bos, J.L. (1988). *New Engl. J. Med.*, **319**, 525.
6. Hadano, S., Watanabe, M., Yokoi, H., Kogi, M., Kondo, I., Tsuchiya, H., Kanazawa, I., Wakasa, K., and Ikeda, J.E. (1991). *Genomics*, **11**, 364.
7. Sarkar, G. and Sommer, S. (1990). *Nature*, **343**, 27.
8. Greer, C.E., Peterson, S.L., Kiviat, N.B., and Manos, M.M. (1991). *Am. J. Clin. Pathol.*, **95**, 117.
9. Shibata, D., Martin, W.J., and Arnheim, N. (1988). *Cancer Res.*, **48**, 4564.
10. Shibata, D., Schaeffer, J., Li, Z.H., Capella, G., and Perucho, M. (1993). *J. Natl Cancer Inst.*, **85**, 1058.
11. Shibata, D., Peinado, M.A., Ionov, Y., Malkhosyan, S., and Perucho, M. (1994). *Nature Genetics*, **6**, 273.
12. Nowell, P.C. (1976). *Science*, **194**, 23.
13. Cavalli-Sforza, L.L., Menozzi, P., and Piazza, A. (1993). *Science*, **259**, 639.
14. Shibata, D., Navidi, W., Salovaara, R., Li, Z.H., and Aaltonen, L.A. (1996) *Nature Medicine*, **2(b)**, 676.

3

In situ hybridization

SHIRLEY A. SOUTHERN and C. SIMON HERRINGTON

1. Introduction

Polymerase chain reaction *in situ* hybridization (PCR ISH) combines components of both PCR and ISH. It is important, therefore, to have some understanding of ISH before attempting its combination with PCR. In direct PCR ISH, the cell/tissue pre-treatment protocols are based on ISH and similar detection systems are used. In indirect PCR ISH (see Chapter 4), PCR is followed by conventional ISH and the procedures are much the same as those used without PCR.

ISH can be used to detect both DNA and RNA by the hybridization of specifically designed nucleic acid probe sequences to intracellular complementary DNA/RNA sequences. Tissue morphology is preserved and therefore ISH has become a valuable tool for precise localization of nucleic acids in intact cells. Non-isotopic methods are now widely used, often in preference to radiolabelled probes, benefiting from shorter procedure times, ease of handling, safety, and equal sensitivity due to the availability of amplification systems for immunodetection.

The main applications of non-isotopic DNA ISH have been for viral detection (1) and interphase cytogenetics (2). Human papillomaviruses (HPVs) have been extensively studied in cervical (3), anal (4), and laryngeal specimens (5). Other viruses analysed include herpes simplex virus (6), cytomegalovirus (7), and hepatitis B virus (8) in various tissues. Interphase cytogenetics allows chromosome-specific sequences to be identified by DNA ISH using repetitive probes (9). Single-copy genes and chromosomal rearrangements can be investigated using yeast artificial chromosome (YAC) and cosmid probes (10).

RNA ISH enables cell-specific mRNAs to be separated from the more abundant mRNAs encoding housekeeping proteins. This allows investigation of gene expression and its relationship to cell differentiation. Detection of low abundance mRNA allows the producer cell to be localized accurately, in contrast with immunohistochemistry where the product may have been rapidly exported from the cell. RNA ISH is also the method of choice for viral identification where viral RNA transcripts are more abundant than the homologous DNA sequence, e.g. Epstein–Barr virus (11).

2. Probes

A wide range of probe types is available for use in ISH including double-stranded DNA, single-stranded RNA (riboprobes), oligonucleotides, and PCR-generated probes. The choice of probe in a particular situation will affect the result (12).

DNA probes are usually doubled-stranded and can be labelled with reporter molecules by nick translation or random primer extension (see Chapter 4). The resultant labelled fragments vary in length (usually 50–500 bp) and, together with the host DNA, require denaturation to single-stranded molecules by heat or chemical means before hybridization. Because of their overlapping segments these probes can produce probe networks or concatenates, amplifying the target sequence. The disadvantage in theory is that these probes can self-anneal during hybridization, resulting in loss of sensitivity due to a decrease in probe available for hybridization. In our hands this has not been a problem provided optimal denaturation and hybridization parameters have been adopted.

Cloned DNA probes can be used to detect several levels of genetic organization. Repetitive sequences are readily detectable, as many such targets (e.g. alphoid, satellite II, and satellite III sequences) are several megabases in length. Smaller probes cloned in YACs or cosmids can also be detected using amplified detection systems. The limit of detection of such an approach using non-fluorescent systems is approximately 10–50 kb and, therefore, it is in the analysis of smaller sequences, such as many single-copy genes and low abundance viral sequences, that PCR ISH is of most potential value (see Chapters 4, 7, and 8).

Riboprobes are produced by cloning cDNA in specifically designed plasmids containing different RNA polymerase promoter sites (e.g. SP6, T7, or T3) in opposite orientation. Sense and anti-sense probes can be produced in this way: anti-sense sequences are complementary to the target mRNA whilst the sense probe acts as a negative control having the same sequence as the mRNA. Labelling is performed by *in vitro* transcription with the proportion of labelled nucleotide altered to give the required sensitivity. Riboprobes are reported to have a hybridization efficiency ten times that of double-stranded DNA probes, which is in part due to the lack of self re-annealment. RNA–RNA hybrids are more stable than DNA–DNA hybrids, allowing the use of higher hybridization temperatures. Moreover, post-hybridization RNase treatment can be used (13) to remove unhybridized probe and reduce background staining.

Oligonucleotide probes are much shorter (usually 20–50 bp) and are commonly prepared by phosphoramidite chemistry. The advantage of short oligoprobes over longer probes is that cell penetration is favoured and specificity is greater. Oligoprobes lack sensitivity, but this can be overcome to some extent by using 'cocktails' of up to 15 oligonucleotides together with 3′-end-labelling systems to add multiple reporter molecules.

PCR can be used to generate labelled probes using specific DNA sequences of interest without the need for cloning. The DNA is denatured and two flanking anti-parallel primers are annealed to the single strands (12). *Taq* polymerase then copies the two single strands by elongation of the primers with insertion of the labelled nucleotide. The amplification reaction is repeated for up to 60 sequential cycles. End-labelled primers can be used as an alternative but the resulting probes contain less label. Primers with RNA polymerase promoter sites can be used to produce templates for *in vitro* transcription.

3. Choice of label

A variety of different isotopic and non-isotopic labels can be incorporated into probes. For ISH, the label must be stable when exposed to the chemicals, solvents, and high temperatures used, and must be easily incorporated into the DNA/RNA using a reproducible labelling system. Low concentrations of the label must be readily detected either directly or following an appropriate detection system. Finally the labelled nucleotide must be designed in such a way that it does not obstruct the labelling reaction and avoids steric hindrance in the subsequent detection system by having a spacer arm of appropriate length.

Originally, non-isotopic labels were thought to be less sensitive than their isotopic counterparts but improved methodology now allows equivalent sensitivity to be achieved with either system, and non-isotopic ISH has the advantages of much higher resolution and less background staining. There are two main types of labelling strategy:

(a) Direct labelling: the detection system is directly attached to the nucleic acid; commonly used direct labels include alkaline phosphatase (AP), horseradish peroxidase (HRP), fluorescein isothiocyanate (FITC), Rhodamine Red (TRITC), Texas Red, aminomethylcoumarin acetic acid (AMCA), Cy 3, and Cy 5.

(b) Indirect labelling: this employs a hapten (e.g. biotin, digoxigenin, mercury, sulfone, or acetylaminofluorene) attached to the probe. The hapten is detected by a labelled binding protein or antibody system.

Of the direct labels, the fluorochromes are popular in molecular cytogenetics, in which it is common practice to use multiple-target ISH (14). Simultaneous detection of several probes can be achieved using FITC, TRITC, and coumarin derivatives (e.g. AMCA). As an indirect label FITC has replaced biotin with some commercially produced probes and, on routine formalin-fixed sections, FITC-labelled probes can be detected with a non-fluorescent system e.g. AP or peroxidase via an anti-FITC antibody to avoid tissue autofluorescence. Cy 3 and Cy 5 are more recently introduced reporter molecules with a fluorescent signal claimed to be ten times stronger than that of FITC.

Biotin and digoxigenin are the most widely used haptens and are attached to the nucleotide via linker arms at optimized lengths to avoid steric hindrance: biotin-11-UTP/dUTP and digoxigenin-11-UTP/dUTP are the most commonly used nucleotide derivatives. At present digoxigenin is often used in preference to biotin as cytoplasmic endogenous biotin present in some tissues, e.g. liver, can complicate interpretation. However, in our hands DNA detection using an anti-biotin detection system (see *Protocol 6*) gives equivalent sensitivity to the digoxigenin systems.

4. Probe labelling

4.1 Nick translation

In this method, labelled nucleotides are introduced into double-stranded DNA by two enzymes: DNase I and DNA polymerase 1 . DNase introduces random nicks along the DNA and the endonuclease activity of the polymerase pulls the labelled nucleotides into place at the 3′-hydroxyl terminus of the nicks. Altering the DNase concentration can affect fragment length but in general probe fragments are between 50 and 500 bp with 60–70% of the label incorporated into the DNA. The overlapping portions of these fragments can anneal together, forming probe networks and hence amplifying the target sequence.

4.2 Random primer extension

Random primer labelling is performed on linearized, denatured DNA to which random oligonucleotides or hexanucleotide primers are annealed. The Klenow fragment of DNA polymerase, which lacks 5′–3′ exonuclease activity, then extends along the template adding the labelled nucleotide. Alteration of the primer concentration can give probes of different fragment sizes and label incorporation of up to 80% can be achieved. This method produces probes with similar properties to nick-translated probes.

4.3 Oligonucleotide tailing

Oligonucleotides can be synthesized chemically to contain a labelled nucleotide at the 5′-terminus or enzymatically tailed at the 3′- or 5′-end of the strand. Using T4 polynucleotide kinase, one label can be added to the 5′-end, generating probes of low sensitivity. Sensitivity can be increased by 3′-end labelling with terminal deoxynucleotidyltransferase (TdT); several labels can be added in this way. Alternatively, a single nucleotide can be added to the 3′-end by using a labelled dideoxynucleotide.

Further details of labelling procedures can be found in Chapter 4 and ref. 12.

4.4 Probe labelling check using nylon membranes

The labelling efficiency of newly produced probes or problematic existing probes can be checked by dot blot hybridization (see *Protocol 1*). This gives a qualitative assessment of the success of labelling. Probe size can be assessed by size fractionation on agarose gels followed by Southern or Northern transfer (15).

Protocol 1. Checking probe labelling

Reagents

- Maleic acid buffer: 100 mM maleic acid, 150 mM NaCl, pH 7.5
- Blocking buffer: 1% non-fat milk diluted in maleic acid buffer
- Tris-buffered saline (TBS): 100 mM Tris–HCl, 50 mM magnesium chloride, 100 mM NaCl, pH 9.0
- Mouse anti-biotin antibody (Dako)
- Mouse anti-digoxigenin AP-conjugated antibody (Boehringer)
- Rabbit anti-mouse antibody (Dako)
- Phosphate-buffered saline (PBS): one PBS tablet (Sigma) in 200 ml water
- Diaminobenzidine (DAB) (see *Protocol 10*)
- NBT/BCIP (see *Protocol 10*)

A. *Detection of digoxigenin-labelled probes*

1. Spot 1 ng of probe on to the membrane and bake for 30 min at 120°C.
2. Wash the membrane in maleic acid buffer for 1 min.
3. Transfer the membrane to blocking solution for 30 min.
4. Incubate in mouse anti-digoxigenin AP-conjugated antibody diluted 1:5000 in blocking solution for 30 min.
5. Wash the membrane twice in maleic acid buffer for 5 min each.
6. Incubate the membrane in TBS pH 9.0 for 5 min.
7. Place in NBT/BCIP substrate until developed (see *Protocol 10*).
8. Wash with water to stop the reaction and allow the membrane to dry.

B. *Detection of biotinylated probes*

1. Follow *Protocol 1A* steps 1–3, then incubate in mouse anti-biotin antibody diluted 1:500 in blocking solution for 30 min.
2. Wash with blocking solution for 10 min.
3. Incubate in rabbit anti-mouse HRP-conjugated antibody diluted 1:400 with blocking solution for 30 min.
4. Wash the membrane twice with maleic acid buffer for 5 min each.
5. Wash with PBS for 5 min.
6. Develop the signal with DAB (see *Protocol 10*).
7. Wash with water and allow to dry.

5. Controls

Non-specific interactions can occur between the probe and unrelated target sequences, or by the probe binding to chemical groups within tissues. In

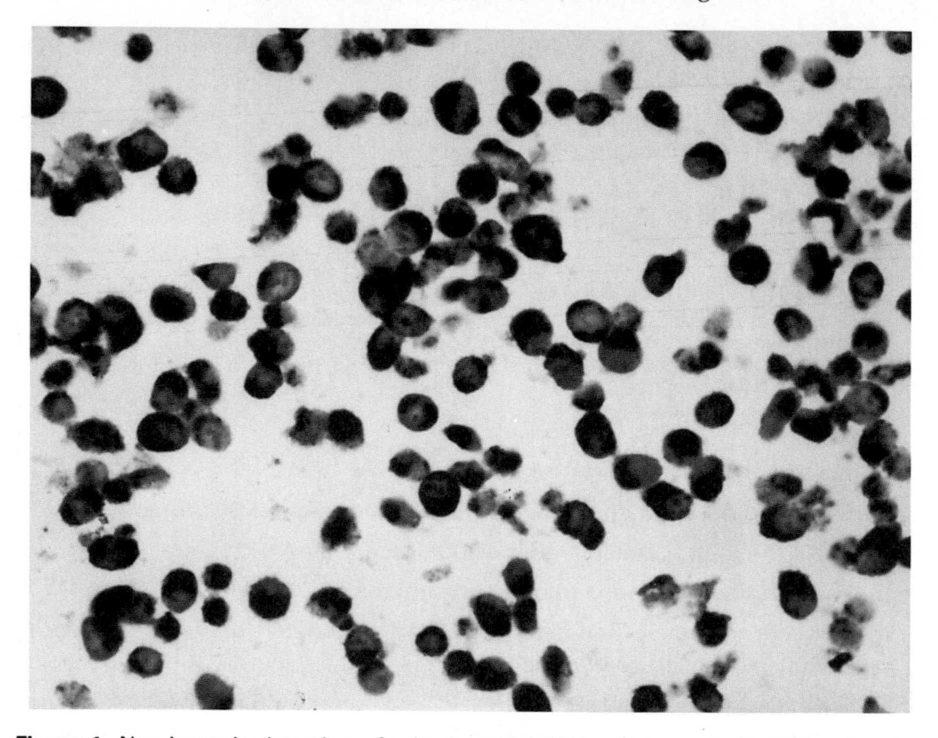

Figure 1. Non-isotopic detection of mitochondrial RNA with a cocktail of digoxigenin-labelled oligonucleotides developed using NBT/BCIP as described in *Protocols 8* and *10*. Note the intense cytoplasmic signal.

addition, simple diffusion of smaller oligoprobes into cells can lead to false or misleading results. Controls must therefore be used initially to establish specificity and sensitivity of the signal and technique, and then continuously included within batches of test slides.

Controls can be divided into a number of categories:

(a) Positive control material: either tissue or a cell line known to have the desired sequence can be used. Cell lines containing low copy numbers are especially useful to establish approximate sensitivity. Paraffin blocks of fixed cell lines may be used (see *Figure 1* and *Protocol 2*).

(b) Positive control probes: mitochondrial RNA probes can be used to check for adequate fixation and mRNA retention in the test samples and to optimize further the technique checking optimal protease digestion and stringency (see *Figure 1*). Total genomic DNA probes indicate adequate denaturation.

(c) Negative control tissue known not to contain the target sequence.

(d) Negative control probes to indicate specificity: sense probes, vector sequence probes or an irrelevant labelled probe are a good indicator of non-target binding either with the probe to an unexpected homologous sequence or to chemical groups present within some tissues.

(e) Digestion of target nucleic acid with RNase or DNase to confirm hybridization to RNA or DNA.

(f) Hybridization with buffer containing no probe, to check for non-specific background caused by the detection reagents or excessive protease digestion. The 'sticky' effect of some hybridization buffer components may, paradoxically, elevate background.

Protocol 2. Preparation of paraffin blocks of control cell lines

Reagents and equipment

- 10% neutral buffered formalin
- PBS (see *Protocol 1*)
- Ethanol (100%, 90%, 75%)
- Xylene
- Wax
- 75 cm^2 flask of cultured cells
- Rubber policeman
- Eppendorf tubes (some of which must be xylene-resistant) (Merck)
- Universal containers

Method

1. Remove the medium from a flask of cells, wash with PBS, and decant the buffer.
2. Add 100 ml of 10% neutral buffered formalin to the flask containing the monolayer of growing cells. Leave the cells to fix for 12–24 h.
3. Tip off the buffered formalin and add 40 ml of PBS, then gently detach the cells from the flask using a rubber policeman.[a]
4. Place the cells/PBS in two Universal containers and centrifuge at 5000*g* for 5 min; discard the supernatant and leave the pellet.
5. Resuspend the cell pellet in 75% ethanol for 30 min, centrifuge, and discard the supernatant.
6. Repeat step 5 once each with 90% ethanol, 100% ethanol, and twice with xylene (at this stage place the cells in a xylene-resistant Eppendorf).
7. Remove the xylene, add sufficient wax to cover the cells, and then incubate at 60°C overnight.
8. Without disturbing the settled cells remove the Eppendorf from the oven and allow the wax to set in the freezer at –20°C.
9. Cut off the end of the Eppendorf containing the cells, carefully remove the conical pellet, and embed upright in paraffin wax.

[a] Use a rubber policeman rather than trypsin to remove the cells, as the latter method can alter digestion times during the ISH method.

6. Specimen collection and fixation

Rapid fixation to avoid loss of nucleic acids and maintain tissue morphology is essential for satisfactory ISH. Tissues are best placed in fixative within 15 min to prevent RNA/DNA degradation by endogenous RNase and DNase enzymes although post-mortem material has been successfully used (16). Cross-linking fixatives, such as formaldehyde, paraformaldehyde, and glutaraldehyde, and precipitating fixatives, including ethanol/acetic acid (17), Bouins (18), Carnoys (19), and Methacarn, have all been used. Different fixatives are only briefly described below as full details are given in Chapter 1.

Precipitating fixatives permit good probe penetration but final tissue morphology is poor and DNA and RNA are lost during the hybridization procedure. Prolonged fixation can lead to extraction or leaching of nucleic acid and should be avoided, but may not be easily controlled in a busy routine laboratory.

Cross-linking fixatives such as buffered formalin and paraformaldehyde are the most suitable for DNA/RNA ISH, with buffered formalin preferable to neutral formalin. The protein lattice retains formed hybrids and, provided the protease digestion is optimized, probe penetration is not a problem. Glutaraldehyde, the best preserver of morphology and nucleic acid retention, is not recommended for ISH as its very extensive cross-linking prevents adequate probe penetration.

7. Preparation of RNase/DNase-free solutions, glassware, and plastics

It is essential to avoid contamination with RNases and DNases which can immediately reduce or abolish the hybridization signal. Gloves should always be worn throughout the procedure including all preliminary chemical handling, solution and section preparation, and the ISH method itself. Where possible, solutions and water should be autoclaved and, for RNA ISH, treated with diethylpyrocarbonate (DEPC) to inactivate RNase (see *Protocol 3*). In our laboratory, however, we find it more convenient and only marginally more expensive to buy most of our stock reagents and chemicals RNase/DNase-free. Preparation with commercially available RNase/DNase free water avoids the toxicity of DEPC, and all the reagents can be freshly prepared just before use.

Glassware, coverslips, and plastics must be autoclaved or baked at 200°C overnight unless too fragile, when treatment with 3% hydrogen peroxide for 10 min is adequate (see *Protocol 3*).

Protocol 3. Preparation of RNase-free solutions, glassware, and plastics

Reagents

- Diethylpyrocarbonate (DEPC) (Sigma)
- Molecular biology grade water (Merck)
- Hydrogen peroxide, 3% (Sigma)

A. *DEPC treatment of solutions*

1. Add DEPC to a final concentration of 0.1% (w/v) to the solution to be treated.[a]
2. Shake well and leave overnight in a fume cupboard.
3. Autoclave the next day.

B. *Preparation of RNase/DNase-free glassware, coverslips, and plastics*

1. Clean glassware thoroughly and rinse in distilled water.
2. Soak glassware, coverslips and plastics in 3% hydrogen peroxide for 10 min.
3. Rinse in molecular biology grade water.
4. Allow to dry and store in a dust-free environment.

[a] Note that Tris buffers cannot be prepared in this way and must be prepared using RNase-free solutes added to DEPC-treated water.

8. Slide adhesive

Due to the harsh treatment of chemicals, proteolytic enzymes, heat treatment and multiple steps involved, it is essential to use a strong adhesive to minimize section loss. A variety of adhesives have been used including poly-L-lysine, gelatin, and albumin but the most effective in our experience is aminopropyltriethoxysilane (APES), which allows the maximum use of digestion enzymes and heat denaturation, a requirement of DNA ISH and even more important for PCR ISH (see *Protocol 4*).

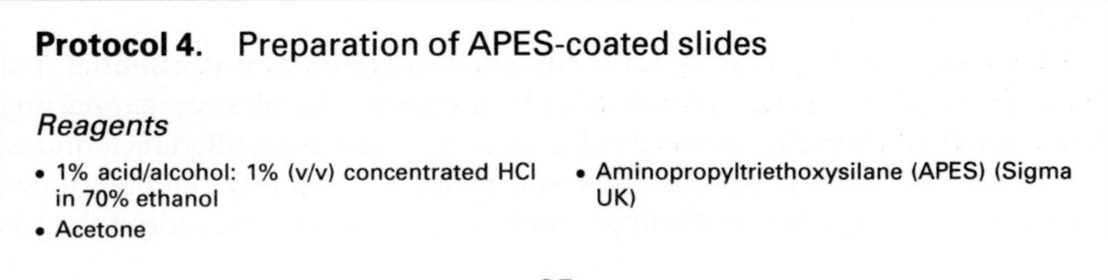

Protocol 4. Preparation of APES-coated slides

Reagents

- 1% acid/alcohol: 1% (v/v) concentrated HCl in 70% ethanol
- Acetone
- Aminopropyltriethoxysilane (APES) (Sigma UK)

Protocol 4. *Continued*

Method

1. Place the slides in a slide rack and clean by immersion in 1% acid/alcohol for 30 min.
2. Rinse thoroughly in running tap water, then wash with distilled water and allow to dry completely.
3. Immerse in acetone for 10 min then transfer to 2% (v/v) APES in acetone for 5 min.[a]
4. Rinse in distilled water twice and then leave to dry at room temperature.
5. Store the slides in a dust-free enviroment and use preferably within one month of preparation.[b]

[a] 800 ml of this solution can be used repeatedly and is sufficient to coat 400 slides but must be used on the day of preparation.
[b] Slides tend to lose adhesiveness with time.

9. Section cutting

Cut sections can be floated out on MBG water within a pre-baked trough in a heated water bath after the contained water reaches the required temperature. Sterile forceps and gloves must be worn to avoid nuclease contamination. The microtome and microtome blades must be cleaned with xylene. Sections are baked on to the APES-coated slides for 30 min at 70 °C or overnight at 60 °C and stored in a dust-free environment until use.

10. Pre-hybridization

10.1 Permeabilization

Aldehyde-induced cross-links between target nucleic acid and other nucleic acids and cellular proteins must to some extent be removed to allow access of probe to the target sequence. Protease enzymes are effective in degrading many of these cross-links and 'unmasking' the nucleic acid. The two most popular enzymes are proteinase K and pepsin but many others are available and used. Proteolysis is the most critical step in the ISH method: too little results in insufficient signal, too much in loss of nucleic acids and morphology. Variation between tissues and blocks, due to the type and extent of fixation, governs the concentration of enzyme which should therefore be determined empirically.

For RNA studies mild protease treatment is required as it is essential that the cell membrane remains intact with adequate cytoplasmic supporting material to sustain the formed hybrid, especially when using oligonucleotides. For DNA ISH nuclear denaturation and probe accessibility requires much harsher proteolysis often to the upper limits of digestion where morphology is just preserved. We find that proteinase K concentrations of <15 μg/ml are optimal for RNA and up to 500 μg/ml for DNA. For ease and consistency proteinase K stock solutions (100 μg/ml for RNA and 5 mg/ml for DNA) can be prepared, aliquoted, and stored frozen at –20°C for up to a year.

Other pre-hybridization steps enhance permeabilization. Mild acid hydrolysis with HCl can increase the hybridization signal. Although the exact mode of action is not clear it may be the result of partial solubilization of nuclear proteins. Detergent solutions such as Triton X and Tween 20 remove membrane lipids and pre-treatment with 2 × SSC at 70°C helps to unravel secondary structures in the target RNA. For long riboprobes it is better to use some of these as extra steps rather than simply to elevate proteinase K concentrations, as introduction of multiple 'types' of permeabilization is preferable to one harsh treatment. Sodium thiocyanate, a powerful protein denaturant, is beneficial in the hybridization of chromosome-specific probes. Some oligonucleotide ISH methods require post-protease fixation with paraformaldehyde (see *Protocol 9*) to stabilize tissue structure.

10.2 Blocking of non-specific probe interactions

Although acetylation of tissues was first introduced when using radiolabelled probes to reduce background staining due to electrostatic forces, non-isotopic ISH can also benefit from this pre-treatment by blocking cytoplasmic amine groups present in tissues, especially material from the gastrointestinal tract (see *Protocol 5*). The signals arising when probes bind to these groups are as strong as or stronger than the specific staining. This is a reported problem of oligonucleotides (20) but we have found that it is also true of longer DNA probes. Acetylation can help with a general 'clean up' of non-isotopic mRNA ISH if the probe binds non-specifically to nuclear material.

Protocol 5. Acetylation of tissue sections

Reagents

- Molecular biology grade (MBG) water
- 1 M triethanolamine stock solution prepared by diluting 22.25 ml triethanolamine in 100 ml of MBG water, adjusting to pH 8.0 with concentrated HCl, then making up to 150 ml with MBG water
- 0.25% acetic anhydride/0.1 M triethanolamine solution prepared by diluting stock 1 M triethanolamine solution 1:10 and, immediately before use, slowly adding 25 μl of acetic anhydride to 10 ml of the diluted triethanolamine.

Method

1. Immediately before applying the hybridization probe (see *Protocol 6*) allow the sections to dry then treat with 0.25% acetic anhydride/0.1 M triethanolamine solution for 10 min under a fume hood.
2. Wash twice with MBG water for 3 min each then allow the sections to dry before proceeding with ISH.

11. Denaturation

For hybridization to occur the probe and target sequences must be single-stranded. This is achieved by denaturation, i.e. the separation of the two strands, either by heat or chemical means using an alkali. Heat denaturation is more commonly used since alkaline denaturation can affect linkages between some reporter molecules and the probe. For DNA ISH, the applied probe and target tissue are generally denatured simultaneously. The temperature must be sufficiently high and the time of exposure to this temperature sufficiently long to separate the DNA strands without compromising tissue morphology. Denaturation temperature and exposure time depend on the extent and nature of fixation and protease digestion; 95–96°C for 6 min is used for DNA viral hybridization (see *Protocol 6*) and other DNA ISH. Single-stranded RNA probes may have secondary or tertiary structures and can benefit from boiling for 5 min then cooling rapidly on ice before hybridization.

12. Hybridization parameters

12.1 T_m

The melting temperature (T_m) is the melting point at which 50% of the probe/target is dissociated. This is important in hybrid formation and stringency control and is influenced by a number of factors:

(a) The GC content of the probe (%G+C): binding between the target nucleic acid and probe is mediated by specific interactions between the complementary A–T and G–C base pairs; there are three hydrogen bonds in G–C base pairs and two in A–T base pairs, so probes rich in G–C sequences are more stable and have a higher T_m.

(b) The probe and target: a hydrophilic hydroxyl group present at the 2′ position of the ribose sugars in RNA stabilizes the double-stranded structure of the duplex. Thus the T_m of RNA–RNA hybrids is significantly higher than for DNA–DNA hybrids.

(c) The length of the probe: the longer the probe the more stable the hybrid.

(d) The composition of the hybridization buffer: hybrid stability is an attribute of hydrophobic interactions between the base pairs. Formamide, an organic solvent, destabilizes these interactions and consequently decreases the T_m of the hybrid, whilst monovalent cations, e.g. Na^+, stabilize the electrostatic forces predominantly caused by the phosphate molecules of the nucleic acid hybrid.

Although the precise numerical value of T_m can be calculated using the equation below, this has been formulated from DNA solution hybridization kinetics and not tissue hybridization and therefore can serve only as a guideline.

$$T_m = 81.5 + 16.6 \log M + 0.41(\%G+C) - 0.72F - 650/L$$

where M is the concentration of monvalent cations, F the concentration of formamide, and L the length of the probe. In practice, optimal conditions are determined empirically.

12.2 Stringency

Hybridization can be carried out under stringent or non-stringent conditions. Stringency refers to the level at which reaction conditions favour duplex dissociation. Under high stringency conditions only duplexes with closely related sequences i.e. high homology are stable. Low stringency conditions allow detection of both closely and distantly related sequences. The desired stringency is determined by temperature, and by formamide and Na^+ ion concentration at the hybridization and post-hybridization washing stages.

In general, temperatures of 20–25°C below T_m with a formamide concentration of 50% and low ionic strength favour a good, fairly fast, stringent hybridization reaction. Temperatures lower than 25°C below T_m and low formamide concentration give low stringency conditions and mis-matching of hybrid base pairs or cross-hybridization can occur. Low stringency methods, for example, allow the use of an HPV-type-specific probe to detect related HPVs.

Inclusion of formamide to decrease T_m allows lower temperatures for hybridization (generally 30–45°C), favouring preservation of tissue morphology and tissue adhesion to the slide.

12.3 Hybridization buffer

Hybridization buffers vary considerably. *Table 1* outlines some components commonly used and their reasons for inclusion in the buffer.

The essential constituents of a hybridization buffer are formamide, dextran sulfate, monovalent cation (Na^+ in SSC), and labelled probe. The other components are a matter of preference. We find for all probe types that simple hybridization buffers containing only the essential constituents give good signals and minimal background. Some workers treat the tissue sections before hybridization with hybridization buffer without probe to equilibrate the tissue. Dehydration before probe application has the same effect.

Probe concentration needs to be determined empirically with a view to achieving a balance between hybrid formation and background noise. As the concentration of probe is increased, the signal increases until a plateau effect or saturation point is reached. After this non-specific background binding is elevated and the true signal is 'masked'. A probe concentration of approximately 1–2 ng/µl, at which target sequences are just saturated, is usually adequate for both DNA and RNA. A volume of 5–20 µl (depending on tissue area) of probe per section is sufficient.

Probe length is important; if the probe is too long, cell penetration is hampered, while short probes are associated with higher background levels

Table 1. Hybridization buffer components

Component	Action
Formamide	Destabilizes hydrogen bonding between probe and target sequences (see Section 12.2)
SSC	Enhances hybrid stability (see Section 12.2)
Dextran sulfate (mol. wt > 8000)	An inert polymer which, by molecular exclusion, concentrates the probe; it also blocks non-specific binding of probe to eosinophils
Sonicated salmon/herring sperm DNA	Blocks non-specific binding of DNA probes
Sonicated yeast tRNA	Blocks non-specific binding of RNA probes
Denhardt's solution containing BSA	Reduces non-specific binding of probe
Ficoll and polyvinylpyrrolidine	Reduces non-specific binding of probe
Tris buffer	Provides a neutral pH
EDTA	Inhibits nucleases
Sodium dodecyl sulphate (SDS)	Reduces background staining

possibly due to adsorption by cells. Despite this a wide range of probe lengths is used, from 20–50 bp for oligonucleotides to 50–500 bp for nick-translated DNA probes and riboprobes.

12.4 Hybridization time

Hybridization time depends on the rate of hybridization which in turn is dependent on hybridization conditions, probe concentration, and target sequences. Provided hybridization conditions are optimal, hybridization time is generally between 1 h and overnight. Abundant target sequences are readily detected after 1–2 h but, for low-copy targets, overnight hybridization may be necessary. Prolonged incubation, i.e. for 2–3 days, does not increase hybridization when using double-stranded DNA probes as probe re-association is favoured in preference to hybridization.

13. Post-hybridization washes

Post-hybridization washes of increasing stringencies can be used to remove unbound components. The stringency of washes depends solely on T_m and is therefore more easily controlled than hybridization stringency, which is influenced by both T_m and hybrid formation. Initial high salt concentration washes (2 $\times$ SSC) remove unbound probe and the subsequent lower ionic strength washes (0.1 $\times$ SSC) at increased temperature remove mis-matched hybrids. Inclusion of formamide is sometimes required for ultimate stringency (21) and specificity, an example being alphoid probes which are only chromosome-specific when 60% formamide washes are used. RNA probes—riboprobes—are prone to non-specific binding because of their 'stickiness' which can be reduced by post-hybridization treatment with RNase (10 μg/ml of

RNase in 2 × SSC for 15 min at 37°C). RNase will selectively remove unbound probe without affecting the RNA–RNA duplex. Oligonucleotide hybrids need gentle treatment and 3% Triton X in PBS or Tris buffer at room temperature will often suffice. Formamide washes can readily cause dissociation of the oligonucleotide probe as can any too stringent conditions with longer probes. Stepwise increase in stringency can be used to assess probe specificity.

14. Detection systems

AP and HRP methods for detection of nucleic acid hybrids are widely used, with AP generally considered to be the more sensitive of the two. Amplification techniques involving multi-layering of specific binding proteins or antibodies have allowed non-isotopic techniques to rival the sensitivity of radiolabelled probes.

Non-isotopic detection can achieve detection of single viral copies (see *Protocol 6*) but such high sensitivity is often achieved at the expense of absolute specificity. Detection of endogenous biotin and non-specific binding of the probe to eosinophils are also enhanced despite heavy use of blocking reagents. The potential use of such a system is the detection of low-abundance viruses and single-copy genes in interphase nuclei.

The streptavidin–biotin system is commonly used because of the high binding affinity of this combination (see *Protocol 9*). Avidin, a glycoprotein present in egg white, has carbohydrate moieties which under physiological conditions may bind to certain lectins owing to their positive charge. It can also bind to negatively charged nucleic acid molecules. Streptavidin produced by *Streptomyces avidinii* does not have carbohydrate moieties so can be used with more specificity to detect biotinylated molecules.

Three different approaches are used with strepavidin/biotin systems; all are standard immunohistochemistry techniques.

(a) Enzyme (e.g. AP or HRP) conjugated to streptavidin in a direct system.
(b) Preformed complex of streptavidin–biotinylated enzyme system.
(c) Streptavidin sandwiched between the biotinylated probe and biotinylated enzyme.

Although biotin is probably the most sensitive reporter molecule, other directly labelled antibodies and their detection systems are available. Digoxigenin, a hapten derived from *Digitalis* (the foxglove), can be visualized via a sheep anti-digoxigenin AP-conjugated antibody (see *Protocol 8*); likewise a FITC-labelled probe can be visualized with an anti-FITC AP-conjugated antibody (see *Protocol 8*). Both systems usually require overnight incubation with Nitroblue Tetrazolium and 5-bromo-4-chloro-3-indolyl-phosphate (NBT/BCIP) AP substrate. These systems are less sensitive and are generally used with oligonucleotides directed against high abundance targets, e.g. infectious viruses. However, any reporter molecule can be enhanced by appropriate antibody layering techniques.

Standard peroxidase methods use an anti-biotin and peroxidase-labelled antibody (see *Protocol 7*) followed by DAB/H_2O_2 peroxidase substrate. The addition of an extra peroxidase-labelled antibody into this method enhances sensitivity if required.

Protocol 6. *In situ* hybridization using biotinylated DNA probes and an alkaline phosphatase detection system

Reagents

- Xylene
- Ethanol (100%, 90%, and 75%)
- Molecular biology grade (MBG) water
- Tris/calcium chloride buffer: 10 mM Tris base, 2 mM calcium chloride, pH 7.4
- Proteinase K (Sigma no. P4914), 500 μg/ml (w/v) in Tris/calcium chloride buffer
- 20 x saline sodium citrate (SSC): 3 M NaCl, 0.3 M sodium citrate, pH 7.0
- 50% (w/v) dextran sulfate prepared by dissolving 5 g dextran sulfate in 10 ml MBG water with overnight incubation at 70°C
- Carrier DNA, e.g. single-stranded DNA (Enzo Diagnostics)
- Streptavidin–alkaline phosphatase (Dako)
- Hybridization buffer prepared by mixing, in a microcentrifuge tube, 50 μl deionized formamide, 20 μl 50% (w/v) dextran sulfate, 10 μl 20 × SSC, 4 μl carrier DNA (10 mg/ml), and 20 μl biotin-labelled probe (5 ng/μl)
- TBS: 50 mM Tris–HCl, 150 mM NaCl, adjusted to pH 7.6 with NaOH
- Buffer 1: TBS containing 2% (w/v) BSA and 0.05% (v/v) Tween 20
- Biotinylated goat anti-avidin antibody (Vector)
- Fast Red substrate (see *Protocol 10*)
- Haematoxylin
- Glycergel (Dako)

Method

1. Place the slides in a metal rack and dewax in xylene for 10 min.
2. Rehydrate the sections by immersing in graded alcohols as follows:
   (a) two changes of absolute ethanol for 5 min each
   (b) 90% ethanol for 5 min
   (c) 75% ethanol for 5 min
   (d) two changes of MBG water for 5 min each
3. Tip off the excess water, place the slides in an incubation tray and allow the sections to dry.
4. Digest the sections with 500 μg/ml of proteinase K for 15 min at 37°C.
5. Wash the slides well with two changes of MBG water for 5 min each.
6. Allow the sections to dry.
7. Add 5–20 μl (depending on the size of tissue) of probe in hybridizing buffer and cover with a cleaned coverslip of appropriate size.
8. Denature the tissue and probe at 96°C for 6 min.
9. Hybridize overnight at 42°C.

10. Remove the coverslip and wash in the following sequence:
    (a) 2 × SSC at room temperature for 30 min.
    (b) 0.1 × SSC at 42 °C for 30 min.
    (c) 0.1 × SSC at 42 °C for 15 min
11. Incubate in buffer 1 for 20 min at room temperature.
12. Incubate with streptavidin–AP conjugate diluted 1:100 in buffer 1 for 30 min at 37 °C.
13. Wash well with buffer 1 for 10 min at room temperature.
14. Incubate with biotinylated goat anti-avidin antibody diluted 1:100 in buffer 1 for 30 min at 37 °C
15. Wash well with buffer 1 for 10 min at room temperature.
16. Repeat step 12.
17. Wash well with TBS at room temperature.
18. Develop with Fast Red AP substrate (see *Protocol 10*).
19. Counterstain lightly with haematoxylin for 15 sec.
20. Mount in Glycergel

Protocol 7. *In situ* hybridization using riboprobes and a peroxidase detection system

Reagents

- Xylene
- Ethanol series (see *Protocol 6*)
- Molecular biology grade (MBG) water
- 0.1% hydrogen peroxide (30% v/v) in methanol
- 0.1 M HCl
- Tris/calcium chloride buffer: 10 mM Tris base containing 2 mM calcium chloride, pH 7.4
- Proteinase K (Sigma), 12.5 μg/ml in Tris/calcium chloride buffer
- TBS: 50 mM Tris–HCl, 150 mM NaCl, pH 7.6 with NaOH
- Mouse anti-biotin monoclonal antibody (Dako)
- Rabbit anti-mouse HRP-conjugated antibody (Dako)
- DAB (see *Protocol 10*)
- Haematoxylin
- DPX
- Hybridization buffer prepared by mixing 50 μl deionized formamide, 20 μl 50% (w/v) dextran sulfate (see *Protocol 6*), 10 μl 20 × SSC (see *Protocol 6*), and 20 μl biotin-labelled riboprobe (5 ng/μl)

Method

1. Dewax the sections in xylene for 10 min.
2. Rinse twice in ethanol for 5 min each.
3. Place in methanol/hydrogen peroxide for 30 min.
4. Rinse twice in MBG water for 5 min each.

Protocol 7. *Continued*

5. Incubate in 0.1 M HCl for 20 min and wash with MBG water.
6. Incubate in 2 × SSC at 70°C for 30 min.
7. Wash with MBG water.
8. Digest with proteinase K solution at 42°C for 15 min.
9. Wash well with MBG water twice for 5 min each wash, drain and allow the sections to dry.
10. Apply 5–20 μl of probe and cover with a coverslip, then hybridize overnight at 42°C.
11. Remove the coverslip and wash with 2 × SSC at room temperature for 30 min.
12. Wash with 0.1 × SSC at 42°C for 1 h.
13. Wash with fresh pre-warmed 0.1 × SSC for a further hour at 42°C.
14. Rinse the sections with TBS.
15. Incubate with mouse anti-biotin diluted 1:40 with TBS for 30 min.
16. Wash with TBS for 10 min.
17. Incubate with rabbit anti-mouse HRP-conjugated antibody diluted 1:50 with TBS for 45 min.
18. Wash with TBS for 10 min.
19. Develop the signal with DAB, controlled microscopically (see *Protocol 10*).
20. Counterstain lightly with haematoxylin.
21. Dehydrate through graded ethanols (75%, 90% then 100%), clear with xylene, and mount with DPX.

Protocol 8. *In situ* hybridization using a digoxigenin- or FITC-labelled oligonucleotide to mRNA and an alkaline phosphatase detection system

Reagents

- Xylene
- Ethanol series (see *Protocol 6*)
- Molecular biology grade (MBG) water
- Tris/calcium chloride buffer (see *Protocol 6*)
- Proteinase K, 12.5 μg/ml in Tris/calcium chloride buffer
- Digoxigenin/FITC-labelled probe
- TBS (see *Protocol 6*)
- Blocking solution: TBS containing 3% BSA, 0.1% Triton X-100, and 20% normal goat serum
- Haematoxylin
- TBS pH 9.0: 100 mM Tris–HCl, 50 mM magnesium chloride, 100 mM NaCl, adjusted to pH 9.0 with NaOH
- Sheep anti-digoxigenin AP-conjugated antibody (Boehringer)
- Rabbit anti-FITC AP-conjugated antibody (Dako)
- NBT/BCIP (see *Protocol 10*)
- Aqueous mountant, e.g. Glycergel
- Hybridization buffer prepared as in *Protocol 7* but with digoxigenin/FITC-labelled probe in place of the riboprobe

Method

1. Dewax the sections in xylene for 10 min.
2. Rehydrate through graded ethanols.
3. Wash with MBG water then allow the sections to dry.
4. Incubate in proteinase K solution for 30 min at 37 °C.
5. Wash well in two changes of MBG water for 5 min each, then allow the sections to dry.
6. Apply 5–20 μl of FITC- or digoxigenin-labelled probe in hybridization buffer (depending on tissue size) and cover with a coverslip.
7. Hybridize at 37 °C for 2 h or overnight if more convenient.
8. Remove the coverslip and wash in two changes of TBS containing 0.1% Triton X-100 for 5 min each.
9. Incubate with blocking solution for 10 min.
10. Tip off the blocking solution and apply rabbit anti-FITC AP-conjugated antibody diluted 1:100 or sheep anti-digoxigenin AP-conjugated antibody diluted 1:500 prepared in blocking solution. Incubate for 30 min.
11. Wash the slides in two changes of TBS for 5 min each.
12. Wash the slides in two changes of TBS pH 9.0 for 5 min each.
13. Develop overnight with NBT/BCIP AP substrate (see *Protocol 10*).
14. Wash the slides in running tap water and counterstain weakly with haematoxylin for 5 sec if required.
15. Wash in running tap water and mount in an aqueous mountant.

Protocol 9. *In situ* hybridization using a biotin-labelled oligonucleotide to mRNA and a streptavidin–biotin AP detection system

Reagents

- Xylene
- Graded ethanols (see *Protocol 6*)
- Molecular biology grade (MBG) water
- Tris/calcium chloride buffer (see *Protocol 6*)
- Proteinase K, 12.5 μg/ml in Tris/calcium chloride buffer
- PBS prepared by dissolving one PBS tablet (Sigma) in 200 ml of MBG water
- 0.4% Paraformaldehyde prepared by dissolving 0.4 g of paraformaldehyde in 100 ml PBS by heating to 70 °C
- TBS (see *Protocol 6*)
- Triton X-100, 0.1% (v/v) in TBS
- Streptavidin–biotinylated AP (Dako)
- Fast Red substrate (see *Protocol 10*)
- Haematoxylin
- Aqueous mountant, e.g. Glycergel
- Hybridization buffer prepared as in *Protocol 7* but with biotin-labelled probe in place of the riboprobe

Protocol 9. *Continued*

Method

1. Dewax the sections in xylene for 10 min.
2. Rehydrate through graded ethanols.
3. Wash with MBG water then allow the sections to dry.
4. Treat with proteinase K solution for 30 min at 37°C.
5. Wash well with MBG water twice for 5 min each wash.
6. Immerse in pre-cooled (4°C) MBG water for 5 min.
7. Immerse in 0.4% paraformaldehyde at 4°C for 20 min.
8. Rinse in two changes of MBG water for 5 min then allow the sections to dry.
9. Apply 5–20 μl of biotin-labelled probe in hybridization buffer and cover with coverslip.
10. Hybridize overnight at 37°C.
11. Remove the coverslip and wash with 2 × SSC twice for 15 min each at 37°C.
12. Wash with 0.1 × SSC for 10 min at 37°C.
13. Treat with TBS containing 0.1% Triton X-100 for 15 min.
14. Incubate with streptavidin diluted 1:100 in TBS for 30 min.
15. Wash with TBS for 10 min.
16. Incubate with biotinylated AP diluted 1:100 in TBS for 30 min.
17. Wash with TBS for 10 min.
18. Develop with Fast Red substate (see *Protocol 10*).
19. Wash with running tap water and counterstain lightly with haematoxylin.
20. Wash with running tap water then mount with aqueous mountant.

15. Substrates

Visualization of the enzyme coupled to the avidin–biotin or antibody molecule can be achieved using a variety of substrates and chromogens (see *Table 2*).

In our hands, the AP reaction with Fast Red and the HRP reaction with DAB are the most appropriate systems for detecting ISH signals. These enzyme reactions are highly sensitive and allow precise localization of the nucleic acid. AP is the more sensitive of the two.

NBT/BCIP is the substrate of choice for overnight incubation as other AP chromogens precipitate non-specifically after extended times. NBT/BCIP gives a more diffuse signal than Fast Red and is not particularly recom-

Table 2. Chromogens used for ISH

Enzyme	Chromogen	Colour
AP	Fast Red TR	Red
AP	Fast Blue	Blue
AP	NBT/BCIP	Blue/black
AP	New Fuchsin	Cerise
HRP	3,3′-Diaminobenzidine (DAB)	Brown
HRP	Hanker Yates	Brown
HRP	3-Amino-9-ethylcarbazole	Brick red
HRP	Tetramethylbenzidine	Blue/black
HRP	4-Chloro-1-naphthol	Blue/purple

mended for precise localization of fine single signals, e.g. demonstration of integrated viral DNA or chromosome identification. It is good, however, for detecting low copy number targets. The AP end-products, except for New Fuchsin, are soluble in organic solvents and therefore require aqueous mountant. If mountants contain high levels of phenol as a preservative, leaching of signal occurs, so mountants with low levels of phenol should be used. Addition of levamisole to the substrate is required to block endogenous AP. This property is used in immunodetection systems where the AP originates from calf intestine and is unaffected by levamisole. In specimens of intestine or where there are high concentrations of placental AP, inactivation is achieved by heating the section in hybridization buffer at 95°C for 15 min before hybridization or by pre-treatment with 20% acetic acid for 15 sec or 0.2 M HCl for 10 min.

Protocol 10. Substrate preparation

A. *Fast Red TR/naphthol AS-MX phosphate substrate (an AP substrate)*

Reagents

- *N,N*-dimethylformamide (DMF) (Sigma)
- Naphthol AS-MX phosphate, free acid (Sigma)
- Fast Red TR salt (Merck)
- Tris buffer pH 8.2: 50 mM Tris–HCl, 150 mM NaCl, adjusted to pH 8.2 with NaOH
- 1 M Levamisole

Method

1. Dissolve 2 mg of naphthol AS-MX phosphate in 0.2 ml of DMF.
2. Make up to 10 ml with Tris buffer then add 10 μl of 1 M levamisole.
3. Immediately before use add 10 mg of Fast Red TR salt and filter.[a]

Protocol 10. *Continued*

4. Cover the sections with substrate and control signal development microscopically.

[a]Substitute Fast Blue BB for a blue reaction

B. *Nitroblue Tetrazolium/5-bromo-4-chloro-3-indolyl phosphate (NBT/BCIP) substrate (an AP substrate)*

Reagents

- Substrate buffer: 100 mM Tris–HCl, pH 9.0, 50 mM magnesium chloride, 100 mM sodium chloride
- 1 M Levamisole
- Nitroblue Tetrazolium (NBT): 75 mg/ml stock solution in 70% DMF
- 5-bromo-4-chloro-3-indolyl phosphate (BCIP): 50 mg/ml stock solution in DMF

Method

1. Prepare substrate solution immediately before use by adding 44 μl of NBT stock solution and 33 μl of BCIP stock solution to 10 ml of substrate buffer.
2. Add 10 μl of levamisole and filter solution.
3. Incubate the sections in substrate solution overnight in the dark.

C. *3,3′-Diaminobenzidine/hydrogen peroxide substrate*

Reagents

- Tris buffer: 50 mM Tris–HCl, 150 mM sodium chloride, pH 7.6
- DAB (Sigma)[a]
- Hydrogen peroxide 30% (v/v)

Method

1. Prepare the substrate solution immediately before use by adding 5 mg of DAB to 10 ml of Tris buffer.
2. Add 50 μl of hydrogen peroxide.
3. Develop signal for approximately 5–15 min, controlled microscopically.

[a]DAB is a suspected carcinogen and must be handled with care.

HRP visualization with DAB is the most popular of the peroxidase methods because of its permanent insoluble end-product. Other peroxidase chromogens are useful for double-labelling ISH. DAB staining intensity can be enhanced using various heavy metals and salts including imidazole, nickel chloride, cobalt chloride, and silver (22), but it is preferable to optimize the ISH rather than rely on end detection enhancement. Endogenous peroxidase present in red blood cells, macrophages, and neutrophils is largely inactivated

Table 3. Troubleshooting

Problem	**Cause(s)**	**Possible solution**
Section floating off during ISH	Loss of APES adhesion due to length of time before use of slides	Prepare fresh slides Section drying time inadequate
Poor morphology	Fixation time too short Overdigestion with protease Over-use of pre-hybridization treatments Denaturation time too long	Correct cause
Weak signal	Inadequate fixation of specimen	Check tissue with a positive control probe e.g. mitochondrial mRNA or genomic DNA probe to assess nucleic acid preservation
	Overfixation or inadequate unmasking preventing probe penetration	Increase protease digestion
	Poor probe labelling	Check label incorporation on a nylon blot (see *Protocol 1*)
	Probe too large	Use alkaline hydrolysis to give a shorter fragments
	Too long an incubation in protease, resulting in target loss	Reduce incubation time
	Loss of hybrid with oligoprobes	Post-fix with paraformaldehyde (see *Protocol 9*)
	Hybridization stringency conditions too high especially with oligoprobes	Try a range of formamide concentrations in the hybridization buffer e.g. 30–50%
	Probe concentration too low	Increase probe concentration
	Weak detection system	Check detection system using a positive primary antibody and immunohistochemistry before proceeding with ISH; include Triton X-100 or Tween 20 in detection system steps
	Nuclease contamination	
High background	Probe concentration too high	Titrate probe
	Antibody concentration too high	Titrate antibody
	Stringency of post-hybridization washes too low	Optimize washes with formamide, salt concentration, and temperature
	Non-specific binding of probe to tissue components	Acetylate tissues; include blocking reagents in hybridization buffer
	Retained unbound probe	Increase hybridization washes; post-hybridization treatment with RNase if appropriate
	Non-specific binding of detection system reagents	Pre-incubate with blocking solution; BSA, normal serum, non-fat milk, and if necessary include in washes and antibody preparations

during DNA denaturation but for RNA ISH treatment with 3% hydrogen peroxide in methanol prior to hybridization will eliminate enzyme activity.

The protocols given here are only selected examples: many other variations exist and are detailed in other volumes in this series (23,24). Many other reporter molecules and detection systems, including fluorescence-based methods, are available, giving great flexibility and sensitivity. Manipulation of appropriate systems enables simultaneous detection and identification of two or three target sequences, which is of particular value when assessing related gene expression during differentiation or embryonic development and chromosome analysis.

16. Troubleshooting

This is detailed in *Table 3*.

References

1. Brigati, D.J., Myerson, D., Leary, J.J., Spalholz, B., Travis, S.Z., Fong, C.K.Y., Hsiung, G.D., and Ward, D.C. (1983). *Virology*, **126**, 32.
2. Herrington, C.S., Cooper, K., and McGee, J.O'D. (1995). *J. Pathol.*, **175**, 283.
3. Cooper, K., Herrington, C.S., Stickland, J.E., Evans, M.F., and McGee, J.O'D. (1991). *J. Clin. Pathol.*, **44**, 990.
4. Wells, M., Robertson, S., Lewis, F., and Dixon, M.F. (1988). *Histopathology*, **12**, 319.
5. Quiney, R.E., Wells, M., Lewis, F.A., Terry, R.M., Michaels, L., and Croft, C.B. (1989). *J. Clin. Pathol.*, **42**, 694.
6. Burns, J., Redfern, D.R.M., Esiri, M.M., and McGee, J.O'D. (1986). *J. Clin. Pathol.*, **39**, 1066.
7. Coates, P.J., Hall, P.A., Butler, M.G., and D'Ardenne, A.J. (1987). *J. Clin. Pathol.*, **40**, 865.
8. Negro, F., Berninger, M., Chiaberge, E., Gugliotta, P., Bussolati, G., Actis, G.C., Rizzetto, M., and Bonino, F. (1985) *J. Med. Virol.*, **15**, 373.
9. Herrington, C.S., Leek, R.D., and McGee, J.O'D. (1995). *J. Pathol.*, **176**, 353.
10. Herrington, C.S. (1994). *J. Histotech.*, **17**, 219.
11. Hamilton-Dutoit, S.J. and Pallesen, G. (1994). *Histopathology*, **25**, 101.
12. Stickland, J.E. (1992). In *Diagnostic molecular pathology: a practical approach*, Volume 2 (ed. C.S. Herrington and J.O'D. McGee), pp. 25–64. IRL Press, Oxford.
13. Waller, A.H. and Savage, K.A. (1994). *J. Histotech.*, **17**, 203.
14. Klinger, K.W. (1995). In *Non-isotopic methods in molecular biology: a practical approach* (ed. E.R. Levy and C.S. Herrington), pp. 25–50. IRL Press, Oxford.
15. Hughes, J.R., Evans, M.F., and Levy, E.R. (1995). In *Non-isotopic methods in molecular biology: a practical approach* (ed. E.R. Levy and C.S. Herrington), pp. 145–182. IRL Press, Oxford.
16. Walker, E. and McNicol, A.M. (1992). *J. Pathol.*, **168**, 67.
17. McCabe, J.T., Morrell, J.I., Ivell, R., Schmale, H., Richter, D., and Pfaff, D.W. (1986). *J. Histochem. Cytochem.*, **34**, 45.

18. Pringle, J.H., Homer, C.E., Warford, A., Kendall, C.H., and Lauder, I. (1987). *Histochem. J.*, **19**, 488.
19. Burns, J., Chan, V.T.W., Jonasson, J.A., Fleming, K.A., Taylor, S., and McGee, J.O'D. (1985). *J. Clin. Pathol*, **38**, 1085.
20. Pagani, A., Cerrato, M., and Bussolati, G. (1993). *Diagn. Mol. Pathol.*, **2**, 125.
21. Herrington, C.S. and McGee, J.O'D. (1994). *Histochem. J.*, **26**, 545.
22. Hsu, S. and Soban, E. (1982). *J. Histochem. Cytochem.*, **30**, 1079.
23. Levy, E.R. and Herrington, C.S. (eds) (1995). *Nonisotopic methods in molecular biology: a practical approach*. IRL Press, Oxford.
24. Williamson, D.G. (ed.) (1996). *In situ hybridization: a practical approach,* 2nd edn. IRL Press, Oxford.

4

PCR *in situ* hybridization (PCR ISH)

JOHN J. O'LEARY

1. Introduction

A major limitation of solution-phase PCR is the inability to visualize and localize amplified product within cellular and tissue specimens. *In situ* hybridization (ISH) does permit localization of specific nucleic acid sequences at the individual cell level. In conventional non-isotopic *in situ* detection systems, most protocols do not detect single-copy genes, except for those incorporating elaborate sandwich detection techniques (1).

Recently, a number of studies have described the use of PCR with ISH (2–6). However, problems are encountered with the technique, with reaction failure or false-positive signals commonly seen. The repertoire of *in situ* amplification techniques has now been extended to include PRINS (primed *in situ* labelling) and cycling PRINS (see ref. 7 and Chapter 6).

2. Definitions and terminology

Five essentially similar *in situ* amplification techniques have now been described:

(a) DNA *in situ* PCR (direct PCR ISH): PCR amplification of cellular DNA sequences in tissue specimens using either a labelled primer or labelled oligonucleotide (dUTP) within the PCR reaction mix. The labelled product is then detected using standard detection techniques as for conventional ISH or immunocytochemistry.

(b) PCR *in situ* hybridization (indirect PCR ISH): PCR amplification of cellular DNA sequences in tissue specimens followed by ISH detection of the amplified product using a labelled internal or genomic probe. The labels used can either be isotopic (^{32}P, ^{35}S) or non-isotopic (e.g. biotin, digoxigenin, or fluorescein). To date, most studies have used non-isotopic labels.

(c) *In situ* reverse transcriptase PCR (*in situ* RT-PCR) (see Chapter 5): amplification of mRNA sequences in cells and tissues specimens by firstly creating a complementary DNA (cDNA) template using reverse transcriptase (RT) and then amplifying the newly created DNA template as for *in situ* PCR.

(d) Reverse transcriptase PCR *in situ* hybridization (RT-PCR ISH) (see Chapter 8): amplification of RNA sequences in cells and tissues specimens by creating a cDNA template using RT, amplifying the newly created DNA template, and then probing this DNA with an internal oligonucleotide probe.

(e) PRINS (see Chapter 6): amplification of specific genetic sequences in metaphase chromosome spreads or interphase nucleic, using one primer to generate single-stranded PCR product. If many rounds of amplification are performed then the technique is called cycling PRINS.

In this chapter I deal with indirect PCR ISH.

3. Principles of the techniques

In situ PCR and PCR ISH techniques represent the coming together of PCR and ISH, allowing the amplification of specific nucleic acid sequences inside cells. Initially, cells or tissues are fixed with a suitable fixative (usually a formaldehyde fixative, e.g. neutral buffered formaldehyde (NBF)) and are then permeabilized using proteolytic enzymes in order to permit access of PCR reagents into the cell and to the target nucleic acid. *In situ* amplification can be performed either in intact cells in Eppendorf tubes (as in solution-phase PCR) or on cytocentrifuge preparations or tissue sections on glass microscope slides.

Many types of equipment have been used for *in situ* amplification techniques, including standard heating blocks, thermal cycling ovens, and specifically designed *in situ* PCR thermocyclers, e.g. GeneAmp *In situ* PCR System 1000 (Perkin–Elmer) and the Omnislide/Omnigene machine (Hybaid) (see Chapter 9). The latter specifically designed machines offer in-built slide calibration temperature curves, which optimize heat transfer kinetics from the thermocycling block to the glass slide to the tissue section.

4. Materials and equipment

Protocols for *in situ* amplification differ greatly between research groups and depend on the target to be amplified and the particular tissue under investigation. Haase *et al.* (2) described *in situ* PCR in fixed single cells, suspended in PCR reaction buffer. After amplification, cells were cytocentrifuged on to glass slides and the amplified product detected using ISH. Other early approaches used pieces of glass slides in standard Eppendorf tubes, incubated directly in PCR reaction buffer with attached cells from cytocentrifuge preparations (8). More recently techniques using tissues and cells attached to microscope slides have been described with amplification carried out either on heating blocks or in cycling ovens (3–5,9). Initially, investigators used standard multi-well PCR blocks. To create an amplification chamber, aluminium foil boats were used into which the slide containing the tissue section was placed. However, thermal conduction was never successfully optimized using standard thermal cycling blocks, with 'thermal lag' (i.e. differences in temperature between the block face, the glass slide, and the PCR reaction mix at each temperature step of the reaction cycle) commonly being encountered (5). The newer *in situ* amplification machines, which offer in-built slide temperature calibration curves, correct for this thermal lag phenomenon.

5. Tissue fixation and preparation

5.1 Cell and tissue fixation

The cytoskeleton of the cell is firstly made rigid in order to create a micro-environment within the cell, which facilitates entry of PCR reagents and minimizes leakage of PCR product. Successful DNA amplification has been achieved with tissues fixed in 1–4% paraformaldehyde, NBF, and 10% formalin for 12–24 h (4,5). Occasionally, tissues fixed with ethanol or acetic acid may be amplified but these always give variable results. Unfortunately, fixation of cells with formaldehyde fixatives has a number of drawbacks. Formaldehyde is not easily removed from tissues, even after tissue processing. Aldehyde groups can react with nucleic acid template to form DNA–DNA and DNA–histone protein cross-links (10). Formaldehyde fixation also allows 'nicks' to occur in the DNA template, some of which are non-blunt-ended and can subsequently act as potential priming sites for extension by *Taq* DNA polymerase; the process occurring even at room temperature. This can lead to spurious results, particularly with *in situ* PCR, when direct incorporation of labelled nucleotide is used.

5.2 Cell and tissue adhesion

Cells must be attached to glass slides in order to prevent them being lost during the repeated cycles of heating and cooling during amplification. Slides pre-treated with coating agents ensure maximal adhesion. The most commonly used coating agents are aminopropyltriethoxysilane (APES), Denhardt's solution, and Elmer's glue.

5.3 Exposure of nucleic acid template

Cell permeabilization is carried out in order to facilitate entry of reagents into the cell. This can be achieved by several methods, including mild protease treatment or mild acid hydrolysis (0.01–0.1 M HCl). In solution-phase PCR extensive proteolytic digestion (occasionally up to 48 h) is employed, to overcome the problem of DNA–DNA and DNA–histone protein cross-linking. However, this cannot be performed with *in situ* PCR techniques, as extensive proteolysis destroys cellular morphology. The maximal digestion time is 20–25 min: longer digestion times compromise architecture. Incomplete dissociation of histone protein–DNA cross-links occurs in *in situ* PCR techniques. Acid hydrolysis probably acts by driving such cross-links to complete dissociation.

6. Amplification: reagents and conditions

Setting up the amplification step, as for solution-phase PCR, involves careful choice of target sequence (taking into account its specific melting temperature, T_m), the choice of primers (again taking into account their T_m, their ability to dimerize, and their uniqueness) and careful optimization of the PCR cycle parameters. Initial denaturation of DNA can be achieved at the beginning of thermal cycling or separately during permeabilization, or, alternatively, following the fixation process. Denaturation can be achieved using heat, heat/formamide, or alkali (11,12). Most investigators advocate the use of hot-start PCR to reduce mis-priming and primer oligomerization. Nuovo (3) has suggested the use of *Escherichia coli* single-strand binding protein, which is involved in DNA replication and repair, to prevent mis-priming and primer oligomerization. The precise mode of action of this protein is unknown.

The volume of PCR reagents varies depending on the size of the cell preparation or tissue specimen being used. In general, between 15 and 75 μl of PCR reagents are used for most *in situ* PCR techniques that use glass-mounted material. Small variations in the volume of these reagents over the slide can give rise to localized PCR failure, contributing to patchy results. When using the GeneAmp *In Situ* PCR 1000 system, 50 μl is recommended. The optimal number of cycles depends on the particular assay. The lower the number of cycles the better, as diffusion of product is less likely to occur and cellular morphology is more likely to be maintained. Most protocols use 25–30 cycles of amplification; exceptionally, 50 cycles may be used. Some groups have used two rounds of 25–30 cycles with the addition of new PCR mixture and *Taq* DNA polymerase, or nested PCR with two rounds of 30-cycle amplification. Primer selection has evolved around two basic strategies: single primer pairs (9–15) or multiple primer pairs with or without complementary tails. The multiple primer pair approach was designed to generate longer and/or overlapping product, with less potential to diffuse from the site of manufacture. In general, single primer pairs suffice if the hot-start PCR modification is adopted.

7. Post-amplification washing and fixation

Most *in situ* amplification protocols include a post-amplification washing step, mainly in order to remove diffused extracellular product and reduce the chances of generating false-positive results. Post-fixation with 4% paraformaldehyde and/or ethanol is also used to maintain localization of product.

8. Visualization of PCR product

Non-isotopic labels provide similar degrees of detection sensitivity to isotopic labels. For ISH after amplification, maximum specificity should only be achieved by probes which recognize sequences internal to the amplified product. However, genomic probes can also be used, which are not restricted to the primer containing sequences and appear to provide similar results.

Post-amplification immunohistochemistry has been described by Nuovo (3), but most investigators find this difficult to perform, because many epitopes do not appear to withstand the repetitive changes of thermal cycling.

9. Reaction, tissue, and detection controls

Judicious use of controls is advised. All reaction set-ups require reference control genes and known negative samples for the target sequence together with irrelevant primers and irrelevant probes. Parallel solution-phase PCR should also be performed for each '*in situ* PCR' assay.

The following are the minimal controls required in PCR ISH:

(a) A reference control gene, e.g. β-globin or pyruvate dehydrogenase (PDH), is important for assessing the degree of amplification in the tissue section or cell preparation; the use of a single-copy mammalian gene such as PDH further allows discrimination of the assay.

(b) DNase digestion: when the target being amplified is DNA, DNase digestion should abolish the signal; if a signal is still present after DNase digestion then the source of the product is either RNA or cDNA or has arisen from spurious amplification.

(c) RNase digestion: in *in situ* RT-PCR, RNase pre-treatment destroys cellular RNA sequences and for DNA assays it reduces the chance of false positives being generated from RNA templates.

(d) Target primers with irrelevant probe: this control is required to assess the specificity of the ISH component of PCR ISH.

(e) Irrelevant primers with target probe.

(f) Irrelevant primers with irrelevant probe: these should not generate a PCR signal from the specific target.

(g) Reference control gene primers with the target probe: this control directly assesses the degree of 'stickiness' of the target probe sequence and the propensity to generate false-positive signal.

(h) Target primer 1 only, and target primer 2 only: using only one primer of the target sequence achieves an asymmetric PCR, with accumulation of minimal product which is not easily detectable.

(i) Excluding *Taq* DNA polymerase: this examines the contribution of primer–primer dimerization and primer oligomerization in generating false-positive signals.

(j) Excluding primers: this examines the role of non-specific elongation by *Taq* DNA polymerase of nicked DNA in tissue sections. This is an extremely important control for DNA *in situ* PCR, when using a labelled nucleotide.

(k) Omitting the RT step in *in situ* RT-PCR or RT-PCR ISH: this is important as only RNA should be amplified if proper DNase digestion has been carried out.

(l) ISH controls for PCR ISH and RT-PCR ISH: to exclude false-positive/negative results due to failure of the ISH step.

(m) Detection controls for immunocytochemical detection systems: to exclude false-positive/negative results due to aberrant staining of the tissues by the detection system.

For *in situ* PCR, controls a, b, c, h, i, and j, together with a set-up including one target primer and one irrelevant primer pair, are used.

10. Problems associated with the techniques

To date, no universally applicable technique is available. The type of starting material, the target, the fixation conditions of the tissues, etc., all appear to influence the reaction. Overall, DNA direct PCR ISH does not work well with material embedded in paraffin wax, because of the problem of non-specific incorporation of nucleotide sequences by *Taq* DNA polymerase. Indirect PCR ISH protocols work if the hot-start modification and/or multiple primer pairs are used. A major advantage of these techniques is their sensitivity, although efficiency is compromised with almost linear amplification achieved. The level of amplification in any particular PCR is difficult to assess. Widely contradictory figures are given: Nuovo (3) reports a 200- to 300-fold amplification, while Embretson (6) estimates amplification of the order of 10- to 30-fold, depending on the number of cycles. Our own observations would tend to support the latter estimation.

The problem of non-specific incorporation of nucleotides by *Taq* DNA polymerase, into damaged DNA, is DNA polymerase- and non-cycle-dependent and occurs frequently in the absence of primers or even with hot-start modification. This makes direct PCR ISH with directly incorporated nucleotides extremely unsatisfactory and generally, I believe, unsuitable for routine use. Gosden (7) has reported the use of strand break joining in *in situ* PCR amplification carried out on chromosomes to eliminate spurious incorporation. Dideoxy blockage during pre-treatment may also minimize spurious incorporation, but on occasion, this is impossible to eliminate. Strand 'super-denaturation', in which double-stranded DNA is denatured at high temperatures and maintained in a denatured state for a prolonged time period (5–10 min), appears to eliminate non-specific incorporation by *Taq* DNA polymerase, but again this is variable in its effect. This 'super-denaturation' routinely precedes conventional hot-start PCR at 70 °C.

10.1 Sequestration of reagents

To carry out successful amplification, increased concentrations (2–5 times higher than solution phase PCR) of reagents are essential. The explanation for using such high concentrations is that some or all the reagents are sequestered during the reaction. Sequestration of reagents can occur by several methods: due to reagents adhering to the glass slides, due to the coating reagents on the slides, or by direct intercalation by the fixative residues present in the tissue. Coating slides with 0.1–1% bovine serum albumin (BSA) allows decreased concentrations of reagents to be used, possibly blocking this sequestration (17).

10.2 Diffusion of amplicons

Diffusion of product from the site of manufacture is encountered. The fewer cycles that are used, the less likely it is that diffusion will occur. Post-fixation with paraformaldehyde and ethanol sometimes helps to maintain localization of the product. Other strategies including overlaying the tissue section with a thin layer of agarose or direct incorporation of biotin-substituted nucleotides (as with *in situ* PCR), which renders the product more bulky and less likely to diffuse, have also been used.

10.3 Patchy amplification

Patchy amplification is always encountered, with on average 30–80% of cells showing amplification signals for the desired target sequence. Patchy digestion during cell permeabilization steps and lingering fixative–DNA and DNA–histone protein interactions all account for this phenomenon. In addition, processed tissues are cut by microtome blades during routine histological preparations, and the tops of many cells are truncated with two immediate effects: firstly, truncated cells may not contain the target sequence, giving rise to a negative result, and secondly, the product is more likely to diffuse from a truncated cell.

11. Current applications

Several groups have worked with PCR ISH techniques and have identified single-copy sequences in cells and low copy number DNA sequences in tissue sections. Many studies have focused on the detection of viral DNA sequences (e.g. human immunodeficiency virus, human papillomavirus (HPV), murine mammary tumour virus provirus, cytomegalovirus, hepatitis B virus, and Kaposi's sarcoma associated herpesvirus) in defined systems. Some groups have examined endogenous human DNA sequences, including single-copy human genes, and chromosomal re-arrangements and translocations (9,13,15). The ability to amplify tumour-specific nucleic acid sequences as possible 'clonal markers' of malignancy (i.e. T-cell receptor gene rearrangements, translocations, point mutations, etc.) has enormous potential for the future, particularly in diagnostic tumour pathology.

12. PCR ISH methodology

12.1 Preparation of slides, cells, and tissue specimens

12.1.1 Preparation of slides

DNA /RNA sequences can be detected in cell smears or in fresh, frozen, or paraffin-embedded sections by ISH or *in situ* amplification techniques. Slides with single rectangular wells (type PH 106 from C.A. Hendley, Essex, UK) or with four round wells (12 mm diameter) (type PH 005, C. A. Hendley) or Perkin–Elmer *in situ* PCR slides are recommended. Hendley slides are best used when performing *in situ* PCR/PCR ISH with standard thermocyclers. With Hendley slides, the wells formed by the thin Teflon coating localize the reagents so that minimal volumes can be used. It is recommended that slides are pre-coated with 3-amino-propyl-triethoxysilane (APES) to maximize section adherence (see *Protocol 1*). Perkin–Elmer slides come pre-treated with silane.

Protocol 1. Coating slides with 3-amino-propyl-triethoxysilane

Reagents

- Decon 90, 2% (v/v) (Sigma): dissolve by stirring 10 ml of Decon 90 in 490 ml of distilled water.
- Acetone, 100%
- 3-Amino-propyl-triethoxysilane (APES), 2% in acetone, prepared by mixing 10 ml of APES in 490 ml of 100% acetone (do this in a fume hood)

Method

1. Immerse the slides in 2% Decon 90 in warm water for 30 min.
2. Wash them thoroughly in distilled water to remove the detergent.
3. In a fume hood, drain off excess water and immerse slides in 100% acetone for 1–2 min.
4. In a fume hood, drain off acetone and immerse in 2% APES in acetone for 5 min.
5. Drain off the excess solution and wash in running water for 1–2 min.
6. Drain off the excess water and allow the slides to dry overnight at room temperature.
7. Store the slides at room temperature until needed.

12.1.2. Preparation of cultured cells

Prepare cultured cells as follows:

(a) Wash cultured cells three times with sterile PBS at 4°C to remove all traces of culture medium.
(b) Re-suspend at 2×10^6 cells per ml and pipette 50 µl on to each well of the pre-coated slides.
(c) Allow the cells to adhere for 10 min, remove any excess, then fix immediately as described in *Protocol 2*.

12.1.2 Preparation of cervical smears

Incubate routine cervical smear slides in methanol/acetic acid (3:1 v/v) for 10 min at room temperature and fix immediately as described in *Protocol 2*. Alternatively, fix the cells in NBF.

12.1.4 Preparation of cryostat sections of fresh frozen tissue

Prepare cryostat sections as follows:

(a) Cut 5 µm sections on to each well of pre-coated slides.
(b) Wash twice in sterile PBS at 4°C to remove all traces of the embedding medium (OCT compound). This material binds many of the detection reagents and normally results in very high background.
(c) Fix sections immediately as described in *Protocol 2*.

12.1.5 Preparation of fixed paraffin-embedded tissues

'Cross-linking' fixatives such as formalin or paraformaldehyde allow reagent penetration only after considerable pre-treatment of the tissue sections. However, despite the harshness of these pre-treatments, the use of these fixatives favours complete preservation of morphology and therefore retention of signal. Prior to paraffin wax embedding, fix the tissue in 10% (v/v) aqueous formalin, 4% (v/v) formaldehyde solution, or NBF. Then proceed as follows:

(a) Cut sections on to slides, and incubate the slides on a hot plate overnight in order to achieve maximum section adhesion. This step prevents loss of sample during the pre-treatment steps.
(b) Dewax the sections by immersing the slides completely in xylene at 37°C for approximately 30 min followed by xylene at room temperature for a further 10 min, and finally immerse the slides in 100% ethanol at room temperature for another 10 min.
(c) Transfer the slides into fresh absolute ethanol before rehydrating through a graded ethanol series to water over a 10 min period.

The slides are then pre-treated as described in *Protocol 4*.

12.2 Tissue and cell pre-treatment

12.2.1 Pre-treatment of fresh culture cells and cryostat sections

Incubation in detergent solutions produces sufficient permeabilization of fresh culture cells and cell smears. A brief incubation in acetic acid can destroy endogenous alkaline phosphatase activity (see *Protocol 2*). This is effective for the removal of liver-type alkaline phosphatase, but usually is less effective in destroying endogenous intestinal and placental phosphatases. Sodium azide/hydrogen peroxide blocks endogenous peroxidase activity. A specialized method for pre-treatment of cervical smears and for use with a peroxidase detection system is given in *Protocol 3*.

Protocol 2. Pre-treatment of fresh cells, cell smears, and cryostat sections

Reagents

- Phosphate-buffered saline (PBS), Sigma
- 4% paraformaldehyde in PBS, prepared by dissolving 12 g of paraformaldehyde in 300 ml of PBS by heating to near boiling point then cooling rapidly on ice prior to use
- 0.1 M Tris–HCl, pH 7.2, prepared by dissolving 24.2 g of Trizma base in 1600 ml of water; adjust the pH to 7.2 with HCl; add distilled water to make up to 2 litres
- 2 × SSC, prepared by diluting 20 × SSC stock solution (20 × SSC: 175.2 g of sodium chloride and 88.2 g of sodium citrate in 1 litre of distilled water; adjust pH to 7.0)
- Acetic acid, 20% (v/v) aqueous solution
- 0.1 M Tris–HCl pH 7.2 containing 0.25% Triton X-100, 0.25% Nonidet P-40: to 1 litre of 0.1 M Tris–HCl pH 7.2 (see above) add 2.5 ml of Triton X-100 and 2.5 ml of Nonidet P-40; warm the solution gently to dissolve the detergents
- Neutral buffered formaldehyde (NBF), prepared by dissolving 4 g of sodium dihydrogen phosphate monohydrate and 6.5 g of anhydrous disodium hydrogen phosphate in 900 ml of distilled water, then adding 100 ml of 40% formaldehyde
- Glycerol, 20% (v/v) aqueous solution

Method

1. Fix the cells in 4% paraformaldehyde in PBS or in NBF for 30 min.
2. Rinse the slides in PBS for 5 min.
3. Immerse the slides in 0.1 M Tris–HCl pH 7.2 containing 0.25% Triton X-100 and Nonidet P-40, twice for 5 min each.[a]
4. Immerse the slides in 0.1 M Tris–HCl, pH 7.2, twice for 5 min each.
5. Immerse the slides in 20% (v/v) aqueous acetic acid at 4°C for 15 sec.
6. Wash the slides in 0.1 M Tris–HCl, pH 7.2, three more times for 5 min each.
7. Incubate the slides in aqueous 20% glycerol for 30 min at room temperature.
8. Rinse with 0.1 M Tris–HCl, pH 7.2.
9. Immerse the slides in 2 × SSC for 10 min.

[a] Additional mild proteolysis may also be required (e.g. 0.1–0.5 mg/ml proteinase K for 15–20 min at 37°C).

Protocol 3. Pre-treatment of cervical smears

Reagents

- PBS (see *Protocol 2*)
- 4% Paraformaldehyde in PBS (see *Protocol 2*)
- PBS–glycine 2% w/v
- 0.1% Sodium azide containing 0.3% hydrogen peroxide
- Proteinase K (Boehringer Mannheim)
- Proteinase K buffer (50 mM Tris–HCl pH 7.6, 5 mM EDTA) prepared by dissolving 6.05 g of Trizma base, 1.86 g of EDTA in 800 ml of distilled water and adjusting to pH 7.6 with HCl; make the final volume up to 1 litre with distilled water

Method

1. Fix the smears in 4% paraformaldehyde in PBS for 15 min at room temperature.
2. Wash in PBS containing 0.2% (w/v) glycine for 5 min.
3. Rinse the slides in PBS.
4. Immerse the slides in 0.1% (w/v) sodium azide containing 0.3% hydrogen peroxide for 10 min.[a]
5. Wash the slides in PBS for 5 min.
6. Incubate the slides in proteinase K solution (1 mg/ml proteinase K in proteinase K buffer) in PBS for 15 min at 37°C.
7. Wash in PBS for 5 min.
8. Immerse the slides in 4% paraformaldehyde or NBF for 5 min.
9. Wash in PBS–glycine for 5 min.
10. Wash in PBS for 5 min.
11. Air dry at 37°C.

[a] This abolishes endogenous peroxidase activity. Care should be taken as this solution is toxic.

12.2.2 Pre-treatment of formalin-fixed, paraffin-embedded tissues

The pre-treatment of formalin-fixed paraffin-embedded sections is described in *Protocol 4* and includes a dilute hydrochloric acid hydrolysis step. With mild acid hydrolysis some limited depurination of nucleic acids occurs. The hydrochloric acid may also partially solubilize the highly cross-linked basic nuclear proteins, found in fixed tissues, and allows easier access of PCR reagents. An incubation with detergent solution alone is insufficient for complete permeabilization of cells. For effective permeabilization, incubation with a protease is essential. Proteinase K is the most effective proteinase for performing this task without destroying overall cellular architecture. The concentration of proteinase K used is critical, if morphology is to be preserved, and is clearly tissue-dependent. The optimum concentration for a particular tissue must be determined empirically by titration, using a positive control for hybridization. Pepsin/HCl digestion can also be used (9.5 ml distilled water, 0.5 ml 2.0 M HCl containing 20 mg pepsin; incubate for 10–30 min at 37 °C).

Protocol 4. Pre-treatment of paraffin-embedded tissue

Reagents

- PBS
- 0.02 M HCl
- Triton X-100, 0.01% in PBS
- Proteinase K[a]
- PBS containing 2 mg/ml of glycine
- Proteinase K buffer (see *Protocol 3*), pre-warmed to 37 °C
- 20% (v/v) aqueous acetic acid
- Graded ethanol series comprising 50%, 95%, and 100% ethanol

A. *Method 1*

1. Immerse the prepared slides in 0.02 M HCl for 10 min.
2. Wash twice in PBS for 5 min each.
3. Extract with 0.01% Triton X-100 in PBS for 3 min (optional).
4. Wash twice in PBS for 5 min each.
5. Equilibrate the slides in pre-warmed proteinase K buffer at 37 °C for 10 min.
6. Incubate the slides with proteinase K solution (proteinase K in proteinase K buffer) at 37 °C for 10–20 min in a Terasaki plate.[a]
7. Wash the slides in two changes of PBS containing 2 mg/ml of glycine for 5 min each.[b]
8. Immerse the slides in aqueous 20% acetic acid at 4 °C for 15 sec to block endogenous alkaline phosphatase activity.[c]
9. Wash in two changes of PBS for 10 min each.
10. Dehydrate through a fresh graded ethanol series comprising 50%, 95%, and 100% ethanol for 5 min each.

[a] The optimum concentration of proteinase K (0.1–5 mg/ml) and the time of digestion will be dependent on the tissue (see text).
[b] This step inhibits the action of the proteinase K and is optional.
[c] When using a peroxidase detection system, peroxidase block is carried out using 0.1% (w/v) sodium azide containing 0.3% hydrogen peroxide. (See *Protocol 3*.)

B. *Method 2*[a]

1. Spot proteinase K in PBS on to the prepared tissue section and place in either a pre-heated oven at 37 °C for 10 min or place into a Terasaki plate floating in a water bath at 37 °C for 15 min.[b]

2. Wash in distilled water
3. Air dry at 75°C.

[a]This is quicker than Method 1, but cellular architecture preservation is not as reliable.
[b]The optimum concentration of proteinase K (0.1–0.5 mg/ml) and the time of digestion will be dependent on the tissue (see text).

C. *Method 3*[a]

1. Follow *Protocol 4A*, steps 1–2.
2. Immerse the slides in 0.01% Triton X-100 in PBS for 90 sec.
3. Rinse the slides in PBS for 2 min.
4. Transfer the slides into a Coplin jar containing 10 μg/ml proteinase K in 0.1 M Tris–HCl, pH 7.5, 5 mM EDTA and place the jar in a microwave oven. Use five slides in 50 ml of buffer per digestion.
5. Microwave the slides in the Coplin jar on full power (800 W) for 12 sec. This should raise the temperature of the solution to 37°C.
6. Leave the Coplin jar in the oven and pulse heat it at 10 min intervals for 5 sec each time at full power. The temperature gradually increases to a level not greater than 50°C.
7. After 30 min, microwave the Coplin jar for 1 min in order to boil the contents.
8. Remove the jar from the oven (**Warning**: take care, as the jar will be extremely hot. Do not cool the Coplin jar, as this will make the glass shatter.)
9. Pour off the hot buffer and transfer the slides to another Coplin jar.
10. Pour on 20% acetic acid at 4°C and incubate the slides for 15 sec.
11. Transfer the slides to PBS and rinse for 2 min.
12. Transfer the slides to fresh molecular biology grade water for 2 min and then dehydrate through graded ethanols to fresh 100% ethanol.
13. Store the slides in 100% ethanol until required.

[a]This method originates from Lewis, F.A. *et al.* (personal communication) and is currently being evaluated.

12.3 Amplification

12.3.1 Amplifying solution

The concentrations of some of the reagents in the amplifying solution for PCR ISH/*in situ* PCR differ from those used for standard PCR. The optimum concentrations are as follows:

- magnesium: 3.5–5.0 mM
- buffer: 10 mM Tris–HCl pH 8.3, 50 mM KCl, 0.001% gelatin

- dNTP: 200 mM
- primers: 1–5 mM
- *Taq* DNA polymerase (*Taq* IS, Perkin–Elmer): 10 units/50 ml

The optimum concentrations of magnesium and *Taq* DNA polymerase are greater than those for standard PCR. This reflects the difficulty of entry of these reagents to the site of DNA amplification and sequestration of magnesium by cellular components and/or non-specific binding of *Taq* DNA polymerase to the glass slide. The amplification procedure is described in *Protocol 5*.

Protocol 5. Amplification

A. *Method 1*

This procedure is based on the Perkin–Elmer GeneAmp *In Situ* PCR Core Kit protocol using the GeneAmp *In Situ* PCR System 1000.

Reagents

- 10 × PCR Buffer II
- 10 mM dATP, dCTP, dGTP, and dTTP
- Ampli*Taq* DNA polymerase IS
- Primers
- 25 mM $MgCl_2$

Method

1. Prepare the reaction mixture by adding the following reagents to 1.5 ml Eppendorf tubes in the proportions given and add water to give a total volume of 49.5 μl
   - 10 × PCR Buffer II, 5 μl
   - dATP, dCTP, dGTP, and dTTP, 1 μl of each
   - Primer 1, 0.5–2.5 μl
   - Primer 2, 0.5–2.5 μl
   - $MgCl_2$, 2–9 μl

   This gives final concentrations of: 1 × PCR Buffer II, 200 mM of each of dATP, dCTP, dGTP, dTTP, 0.2–1 mM of each primer, and 1–4.5 mM $MgCl_2$.
2. Turn on the Perkin–Elmer GeneAmp *In Situ* PCR System 1000 and start a SOAK file at 70°C.
3. Plug in the Assembly Tool. Turn on a separate heat block and allow it to reach 70°C.
4. Put the reaction mix tubes into the 70°C heat block and incubate for 5 min.
5. After the 5 min incubation, add the appropriate amount of Ampli*Taq* DNA polymerase IS (0.5 μl per 50 μl total volume) to each reaction

mix tube, mix gently, and return the tube to the heat block. If different tubes differ in their primer composition, be careful not to cross-contaminate the mixtures.

6. Lay an AmpliCover Clip on a clean surface so that the oval recess is face up.

7. Gently slide the grips to the open position and place an AmpliCover Disc, with the tabs down, into the recess of the AmpliCover Clip. Try to avoid pushing the centre of the AmpliCover Disc.

8. Repeat this step until enough AmpliCover Clip/AmpliCover Disc assemblies have been prepared to accommodate all of the samples in the experiment.

9. Place a prepared slide on the Assembly Tool platform surface, positioning the slide so that the first sample is over the inscribed oval on the platform. The samples must be loaded consecutively, starting with the sample furthest away from the frosted end of the slide.

10. Place an AmpliCover Clip and AmpliCover Disc assembly into the magnetic slot in the Assembly Tool arm. Make sure the assembly is correctly aligned in the slot and the grips are in the open position.

11. Dispense 50 μl of the heated reaction mix directly on to the sample spot, being careful not to touch the sample with the pipette tip.[a]

12. Lower the arm of the Assembly Tool and press gently until it latches. Slowly turn the handle with a smooth motion until it stops. Do not reverse direction after the AmpliCover Disc has made contact with the reaction mixture.

13. Press the two levers on top of the Assembly Tool arm to engage the grips on to the edges of the slide.

14. Release the levers, disengage the handle, and raise the arm of the tool.

15. Move the slide along the platform surface until the second sample is in the loading position and repeat steps 10–14. Finally, load the third sample.

16. Grip the slide by the edges of the frosted label area and lift it from the platform surface.

17. Open the sample block cover of the pre-heated (70°C) instrument and raise the slide retaining lever.

18. Place the slide into one of the vertical slots of the sample block. This slide can only fit in one orientation (see *User Manual* for the Perkin–Elmer GeneAmp *In situ* PCR System 1000 for further details).

19. Lower the retaining lever and close the block cover.

20. Repeat the loading sequence with the remaining slides.

Protocol 5. *Continued*

21. Commence thermal cycling using a typical cyling programme of: 94°C for 1 min followed by 94°C for 55 sec; 55°C for 45 sec × 30 rounds then cool the slides at 4°C.[b]

[a] Pipette gently, forming a bead of reaction mix over the sample. Pull the pipette tip away from the slide while delivering the solution to form as tall a bead as possible. Avoid forming bubbles on the bead surface by not expelling air from the pipette.
[b] For *in situ* PCR with labelled primers, do not allow the slide to cool to 4°C, as this will lead to spurious binding of labelled primer to cells and tissues. Ideally, the temperature should only be allowed to cool to 5°C below the predicted T_m of the primer.

B. *Method 2*

This procedure uses a standard thermal cycler. It can be simplified by following the dewaxing steps in the protocols from Section 12.1.5 and then by proceeding directly to the proteinase K digestion (see *Protocol 4*).

Equipment and reagents

- PCR buffer: 50 mM KCl, 10 mM Tris–HCl pH 8.3 at room temperature, 4.5 mM $MgCl_2$, 0.01% gelatin, 200 mM of each dNTP, and 5 mM primer
- Taq DNA polymerase (2.5 units per 12.5 μl)
- Nail varnish
- Mineral oil, pre-heated to 80°C
- Chloroform
- Absolute ethanol
- Aluminium foil
- Gel bond (FMC Bioproducts)
- Thermal cycler (480 DNA thermal cycler, Perkin–Elmer)

Method

1. Place the slide containing the fixed cell suspension in an aluminium foil boat, trimmed to slightly larger proportions than the slide.
2. Place 10 μl of PCR buffer on top of the cell suspension contained in the deep well slide.
3. Cover the well with a pre-cut piece of Gel bond, hydrophobic side down.
4. Place the slide (in the aluminium foil boat) on the heating block of the thermal cycler and allow the temperature to increase to 80°C.
5. Once the temperature has reached 80°C, lift a corner of the Gel bond and add 2.5 μl of *Taq* DNA polymerase to the PCR mix contained on the slide.
6. Replace the coverslip and seal the margins with nail varnish.
7. Place 1–2 ml of preheated mineral oil on top of the slide to ensure optimum thermal kinetics.
8. Apply the following PCR protocol: 94°C for 6 min followed by 40 cycles of 55°C for 2 min and 94°C for 1 min.
9. Following amplification, dip the slide in chloroform to remove

mineral oil and carefully remove the Gel bond coverslip. As the Gel bond hydrophobic side is towards the cell suspension, the PCR reagent mix remains on the tissue in the well.

10. Carefully dip the slides in 100% ethanol and dehydrate.[a]
11. Proceed with the hybridization of biotin- or digoxigenin-labelled probe to the amplified DNA.

[a] The slides can also be post-fixed in 2% paraformaldehyde to maintain localization of PCR product if required.

12.4 *In situ* hybridization

(See also Chapter 3.)

12.4.1 Selection of probes and probe labels for PCR ISH

Recombinant DNA technology now provides the opportunity to use DNA or RNA probes to any desired sequence. Furthermore, one can choose between single-stranded and double-stranded probes. The type of probe to be used depends greatly on the target sequence. Probes to DNA viruses are constructed as recombinant plasmids or cosmids, with whole viral DNA or restriction enzyme fragments of specific viral sequences. Many viral and human probes are now commercially available, and may be obtained from the originating laboratory or from the American Type Culture Collection. Double-stranded probes, if randomly sheared, form networks on the cytological hybrid and so increase the hybridization signal. However, double-stranded probes can also anneal in solution and thus reduce the concentration of probe available for reaction with the cytological preparation. Recently, human and viral oligonucleotide probes have been constructed synthetically using oligonucleotide synthesizers. These oligonucleotides, containing sequences complementary to DNA or RNA viruses or human sequences, can easily be labelled and then used as probes. Their use in conjunction with non-isotopic detection systems has been limited because the short sequences of these probes allow less room for adequate labelling. To use such oligonucleotides successfully in non-isotopic methods, it is necessary to use cocktails containing up to four different probe sequences.

The general use of ISH is hindered by problems associated with probe preparations. Initially probes were prepared using crude isolated nucleic acids. In addition, only cumbersome, expensive, and dangerous radioactive labels could be used. These difficulties have been largely overcome by recombinant DNA technology. The development of a biotinylated nucleotide analogue, biotin-11-dUTP (Enzo Diagnostics Inc.) and the development of DNA labelling, by nick translation, greatly facilitated the use of non-isotopic ISH. The use of biotinylated viral probes was initially described using immunocytochemical detection, but the development of a technique using a biotin/streptavidin/alkaline phosphatase sandwich detection system proved to be more sensitive. Subsequently, multiple sandwich immunohistochemical techniques have recently become available(3). Problems exist when using biotin, as many tissues, including liver, small bowel, and endometrium, have high levels of endogenous biotin, which then interferes with the detection of the biotinylated hybridization products. Alternative molecules used include aminoacetylfluorene- and mercury-labelled probes that allow the detection of nucleic acid sequences in fresh isolates or cultured cells by fluorescence methods. Generally, fluorescence detection systems are not readily applicable to archival, paraffin-embedded sections because of endogenous autofluorescence. However, digoxigenin, a derivative of the cardiac glycoside digoxin, conjugated to dUTP is commonly exploited. Most DNA probes (especially viral ones) contain long double-stranded sequences and consequently nick translation is the most general and effective method for labelling probes with either biotin or digoxigenin. Nick translation has the advantage of producing large quantities of labelled probe that can be stored readily at –20°C for up to two years without any significant deterioration, and gives probes of optimum size, which range from 300 to 800 bp. Labelling an entire recombinant plasmid, so the probe can be used with 'built-in carrier DNA', enhances signal strength. Random primer labelling methods yield low levels of product (70 ng) and require pre-digestion of the plasmid giving excised fragments that include the viral sequence. Of the biotin deoxynucleotides available, biotin-11-dUTP allows optimum detection sensitivity. Separation by spun column purification removes unincorporated nucleotides from labelled probes and therefore reduces high levels of background staining.

Methods for nick translation and random primed labelling of genomic probes are given in *Protocols 6* and *7*. 3′-End-labelling is described in *Protocol 8*; probes labelled in this way can be further labelled at the 5′-end by synthesizing them with a 5′ aliphatic amino group which is free for conjugation with several different fluorochromes. Alternatively, alkaline phosphatase can be coupled directly to short synthetic oligonucleotides. These have single modified bases containing a reactive amino groups that enable covalent cross-linking.

Protocol 6. Nick translation[a]

Reagents

- 10 × nick-translation buffer (10 × NT buffer): 0.5 M Tris–HCl pH 7.8, 50 mM $MgCl_2$, 0.5 mg/ml BSA (nuclease-free)
- 10 × dNTPs: 0.5 mM dATP, dGTP, dCTP, dTTP with one triphosphate replaced by the labelled nucleotide at optimum concentration, e.g. DIG–dUTP replaces dTTP
- Dithiothreitol (DTT), 100 mM
- DNase I, 1 mg/ml
- Digoxigenin-11-dUTP (DIG-dUTP) as the labelled triphosphate (Boehringer-Mannheim)
- DNA polymerase I, 5 U/ml
- Glycogen, 20 mg/ml
- EDTA, 0.5 M, pH 8.0
- 3 M Sodium acetate, pH 5.6
- Sephadex G-50 column
- Carrier herring sperm DNA
- Tris–EDTA buffer (TE): 10 mM Tris–HCl, pH 8.0, 1 mM EDTA
- Ethanol (–20°C)

Method

1. Add the following to one 1.5 ml Eppendorf tube:
   - DNA, 1 μg
   - 10 × NT buffer, 5 μl
   - 10 × dNTPs, 5 μl
   - 0.5 mM labelled triphosphate, 5 μl
   - 100 mM DTT, 5 μl
   - 1 mg/ml DNase I, 5 μl
   - DNA polymerase I, 5 μl

   Make up the final volume to 50 μl with distilled water.
2. Mix and incubate at 15°C for 2 h.
3. Purify the labelled probe by Sephadex G-50 chromatography (column previously equilibrated with TE).
4. Add 1 μl glycogen or carrier DNA to the eluate and precipitate with ethanol by adding 0.1 volume of 3 M sodium acetate, and 2.5 volumes of absolute ethanol.
5. Incubate at –20°C for 1 h and centrifuge at 15 000*g* for 30 min at 4°C.
6. Dry the pellet and resuspend in TE.

[a] This protocol is adapted from ref. 18.

Protocol 7. Random primed labelling

This protocol is adapted from ref. 18 and is performed on a linearized, single-stranded DNA template. The oligonucleotides are annealed randomly and then conjugated with labelled nucleotides. By comparison with nick translation, the probes labelled by random priming are composed of a larger portion of small fragments giving a greater incorporation of the label.

Reagents

- DNA
- 10 × hexanucleotide mixture (10 × HM): 0.5 M Tris–HCl, 0.1 M $MgCl_2$, 1 mM dithioerythritol, 2 mg/ml BSA, 62.5 A_{260} U/ml hexanucleotides, pH 7.2
- 10 × dNTPs: 0.5 mM dATP, dGTP, dCTP, dTTP with one triphosphate replaced by the labelled nucleotide at optimum concentration, e.g. DIG–dUTP replaces dTTP
- Klenow enzyme (2 U/μl)
- Digoxigenin-11-dUTP (DIG-dUTP) as the labelled triphosphate (Boehringer Mannheim)
- 0.5 M EDTA, pH 7.4
- 3 M Sodium acetate, pH 5.6
- Sephadex G-50 column
- Carrier herring sperm DNA
- TE: 10 mM Tris–HCl, pH 8.0, 1 mM EDTA
- Ethanol (–20°C)

Method

1. Linearize and denature 10 ng–3 μg of DNA by heating for 10 min at 95°C then chill on ice for 5 min.

Protocol 7. *Continued*

2. Add the following to a single 1.5 ml Eppendorf on ice: 2 μl of 10 × HM, 2 μl of 10 × dNTPs, 2 μl of labelled triphosphates (0.5 mM), 10 ng–3 μg of DNA; make up the final volume to 19 μl with distilled water.
3. Add 1 μl (2 U) Klenow enzyme.
4. Mix, micro-centrifuge briefly at 15 000*g*, and incubate for at least 60 min at 37 °C.
5. Stop the polymerase reaction by adding 2 μl 0.5 M EDTA, pH 7.4.
6. Precipitate the labelled DNA by adding 0.1 volume of 3 M sodium acetate and 2.5 volumes of ethanol.
7. Keep at –20 °C for 2 h and centrifuge at 15 000*g* for 30 min at 4 °C.
8. Wash the pellet with cold ethanol, carefully remove the solvent, and dissolve the pellet in TE buffer. Store at –20 °C until use.

Protocol 8. 3′-End-labelling of oligonucleotide probes[a]

Reagents

- DNA
- 10 × terminal transferase buffer (10 × TTB): 1 M potassium cacodylate, 10 mM $CoCl_2$, 250 mM Tris–HCl, 2 mM DTT, pH 7.6, 2 mg/ml BSA
- 10 mM dATP
- Digoxigenin-11-dUTP (DIG-dUTP) as the labelled triphosphate (Boehringer Mannheim)
- Oligonucleotide
- Terminal deoxynucleotidyltransferase (TdT) Boehringer Mannheim
- 4 M LiCl
- Ethanol (–20 °C)
- 0.2 M EDTA, pH 8.0
- TED buffer: 10 mM Tris–HCl, pH 8.0, 1 mM EDTA, 10 mM DTT

Method

1. Add the following to a 1.5 ml Eppendorf tube:
   - 100 pmol of oligonucleotide
   - 2 μl of 10 × TTB
   - 1 μl of dATP
   - 1 μl of labelled triphosphate
   - 1 μl of terminal transferase (25–50 U)

   Make up the final volume to 20 μl with distilled water.
2. Incubate for 1 h at 37 °C.
3. Stop the polymerase reaction by adding 2 μl of 0.2 M EDTA, pH 8.0.
4. Precipitate the labelled DNA by adding 2.5 μl 4 M LiCl and 75 μl of –20 °C ethanol.
5. Keep at –20 °C for 2 h and centrifuge at 15 000*g* for 30 min at 4 °C.

6. Wash the pellet with cold ethanol, carefully remove the supernatant, and dissolve the pellet in TED buffer. Store at –20°C until use.

[a] Adapted from ref. 18.

12.4.2 Hybridization

Genomic probes

Biotinylated and digoxigenin-labelled or fluorescent genomic DNA probes are usually prepared at a concentration of 200 ng/ml in one of the hybridization buffers described in *Protocol 9*. It is usually best to use the simplest hybridization buffer possible and this is usually adequate for producing successful hybridization at temperatures of 37–42°C with overnight incubation. Hybridization buffer 2, *Protocol 9*, is an example of this. The milk powder present in this buffer successfully blocks non-specific probe-binding sites, which obviates the need for complex mixtures containing phenol, BSA, and polyvinylpyrrolidone. There appears to be no significant advantage in incorporating a carrier DNA, such as single-stranded DNA (e.g. salmon sperm) into hybridization mixtures.

Hybridization buffers that contain dextran sulfate are very viscous. This usually produces a high surface tension on contact with a glass surface. Therefore, if glass coverslips are used to cover tissue sections during hybridization reactions, they are subsequently difficult to remove as the surface tension which is induced between the glass and the buffer produces enough suction to lift the section from the slide. For this reason, I recommend the use of Gel bond film (FMC Corporation), cut to coverslip size to cover the sections. This material is a pliable plastic and can therefore easily be removed following hybridization. It also has one hydrophilic and one hydrophobic surface; it is usually placed with the hydrophobic surface facing the tissue section, which results in no surface tension with the buffer. Gel bond is sealed in place with nail varnish to prevent leakage out of probe and leakage in of moisture during hybridization. Alternatively, rubber cement or agarose can be used.

Before hybridization (either overnight or at 4 h) at the desired temperature, both cellular DNA and probe DNA (if double-stranded) must be simultaneously denatured by heating the slides to 90–95°C for 10 min. This is accomplished by placing the slides on a preheated baking tray in an oven at the desired temperature. The temperature and timing of this step are critical, if tissue morphology is to be preserved. Incubation of the slides at 37–42°C overnight results in high hybridization efficiency, particularly with DNA virus probes, and is stringent enough to produce specific hybridization with little or no cross-hybridization between viral types.

Oligonucleotide probes

Probe concentrations of 5–10 pmol/100 μl are advised. The concentration depends on the type of labelling method employed. When using 'tailed' probes, lower concentrations are required. Hybridization buffer 4 (see *Protocol 9C*) is suitable for most applications. Conditions of hybridization are in general similar to those used for genomic probes. Ideally, the T_m of the probe should be calculated. However, hybridization kinetics within tissue sections do not follow classical T_m solution hybridization kinetics.

Protocol 9. Hybridization of DNA probes to pre-treated tissue sections

A. *Preparation of hybridization buffer 1*[a] *(suitable for genomic probes)*

Reagents

- Dextran sulfate
- 200 mM Tris–HCl pH 7.2
- Ficoll
- Polyvinylpyrrolidone
- BSA
- NaCl
- Sodium citrate
- Single-stranded salmon sperm DNA
- Formamide

Method

1. Add 20 g of dextran sulfate to 25 ml of 200 mM Tris–HCl pH 7.2.
2. Heat with stirring to dissolve.
3. Dissolve separately 0.04 g each of Ficoll, polyvinylpyrrolidone, and BSA in 5 ml of 200 mM Tris–HCl pH 7.2 by heating.
4. Allow all the solutions to cool to room temperature and combine them with the cooled dextran sulfate solution (giving dextran sulfate/Denhardt's solution).
5. Add 3.5 g sodium chloride and 1.76 g sodium citrate to 10 ml of 200 mM Tris–HCl pH 7.2.
6. Stir to dissolve and combine with the dextran sulfate/Denhardt's solution.
7. Add 8 ml of a 62.5 mg/ml solution of single-stranded salmon sperm DNA in water.
8. Adjust the total volume of the buffer to 100 ml with distilled water.
9. Store this double-strength buffer at 4°C.
10. For hybridization, take the stock buffer and add an equal volume of formamide to give buffer containing 50% formamide.

B. *Preparation of hybridization buffer 2 ('Blotto buffer')*[b] *(suitable for genomic probes)*

Reagents

- Dried milk powder (pure, containing no vegetable extracts)
- Dextran sulfate
- 20 × SSC (see *Protocol 2*)
- Formamide

Method

1. Dissolve 0.4 g of dried milk powder in 10 ml of distilled water.
2. Add 10 g of dextran sulfate to 50 ml distilled water and stir to dissolve (heat gently if necessary).
3. When cool add 20 ml of 20 × SSC and the milk solution and make up to a final volume of 100 ml with distilled water.
4. Add an equal volume of formamide to give working strength Blotto buffer.

C. *Preparation of Hybridization buffer 3*

Reagents

- 50% Formamide
- 5% dextran sulphate
- 2XSSC
- 50mMTris-HCl, ph 7.4
- 0.1% (w/v) sodium pyrophosphate
- 0.2% (w/v) polyvinyl pyrrolidone (mol wt 40000)
- 0.2% (w/v) Ficoll 9mol. wt 400000)
- 5mM EDTA
- 200ng/m sheared human DNA

This buffer is particularly useful when using cocktails of genomic or oligo probes.

Method

1. Dissolve 20 g of dextran sulfate in 25 ml of 200 mM Tris HCl pH 7.4.
2. Heat with stirring until dissolved.
3. Dissolve separately 0.4 g of each of Ficoll and polyvinylpyrrolidone, and 0.2 g sodium pyrophosphate, in 5 ml of 200 mM Tris HCl pH 7.4 by heating.
4. Allow the solutions to cool and combine them with the cooled dextran sulfate solution.
5. Add 3.5 g sodium chloride and 1.76 g sodium citrate to 10 ml of 200 mM Tris HCl pH 7.4.
6. Stir to dissolve and combine with the solution from Step 4.
7. Add 2 μl of a 10 mg/ml solution of sheared human DNA.
8. Adjust the total volume of the solution to 100 ml with distilled water.
9. Store at 4°C.
10. For hybridization, add an equal volume of formamide to give buffer containing 50% formamide.

D. *Preparation of buffer 4 (for oligoprobes)*

Reagents

- 2XSSC
- 5% dextran sulphate
- 0.2% dried milk powder

Protocol 9. *Continued*

Method

1. Follow steps 1–3 of *Protocol 9B.*
2. For hybridization, add an equal volume of distilled water to give the basic buffer.
3. Alternatively, if formamide is required[d], add 0.8 volumes of distilled water and 0.2 volumes of formamide to the solution from Step 1 to give a 10% formamide, 2 × SSC, 5% dextran sulfate, 0.2% (w/v) dried milk powder solution.

E. *Hybridization*

Equipment and reagents

- Hybridizaton buffer: buffer 1 (see *Protocol 9A*), buffer 2 (see *Protocol 9B*), buffer 3 (see *Protocol 9C*), or buffer 4 (see *Protocol 9D*)
- Probe
- Gel bond film (FMC Bioproducts)
- Nail varnish
- Oven set at 90–95°C and containing pre-heated baking tray
- Humidified box

Method

1. Re-consititute the probe in the appropriate hybridization buffer; the optimum concentration varies for different slides and types of probes (see step 2).
2. For Perkin–Elmer slides, apply 30 μl of genomic probe (2 ng/μl) to each section. For PH106 C.A. Hendley slides, apply 75 μl of the appropriate genomic probe (200 ng/ml) to each section. For multi-well slides (PH005, C.A. Hendley), apply 8–10 μl of genomic probe (2 ng/μl). For oligonucleotide probes apply 20–30 μl of probe (5–10 pmol/100 μl). For cervical smears, use hybridization buffer 3[c] (above) and a cocktail of probes (HPV 6, 11, 16, 18, 31, 33) at a concentration of 2 ng/μl of each probe.
3. Cut Gel bond films to coverslip size and place, hydrophobic side downwards, over each section. Seal the Gel bond in place with clear nail varnish (for multiwell slides, it is advisable not to seal with nail varnish: a plastic coverslip may be used).
4. Place the slide on to a pre-heated baking tray and incubate at 90–95°C for 10 min.
5. Transfer slides to a humidified box and incubate slides at 37–42°C overnight or at 42°C for 2 h.

[a]This makes a double-strength buffer. The composition of 1 × hybridization buffer 1 is 10% dextran sulfate, 2 × SSC, 2 × Denhardt's solution (0.02% Ficoll, 0.02% polyvinylpyrrolidone, 0.02% BSA, 5 mg/ml single-stranded salmon sperm DNA) in Tris–HCl pH 7.2, 50% formamide.

[b]The composition of the working solution is 2 × SSC, 5% dextran sulfate, 0.2% (w/v) dried milk powder, 50% formamide. This hybridization buffer is easily made and gives results comparable to or better than buffer 1.
[c]This buffer is particularly useful when cocktails of genomic probes are used for the analysis of routine cervical smears for the detection of HPV 6, 11, 16, 18, 31, 33, etc.
[d]The requirement for formamide is determined by experiment using appropriate controls.

12.5 Post-hybridization washing

Post-hybridization washes are performed most efficiently with agitation. This can be accomplished simply by using an orbital shaker or inverting a Coplin jar lid into the bottom of a glass staining dish and placing a magnetic follower in the upturned lid. Fill the dish with wash buffer and immerse the staining rack into the buffer so that it rests on the edge of the Coplin jar lid. The buffer can now be stirred vigorously on a magnetic stirrer without interference with the magnet's motion.

12.5.1 Washing with increasing stringency

Genomic probes

The object of the washing protocol is to remove excess or non-specifically bound probe and any mis-matched hybrids that may have formed during hybridization. To achieve this, washing protocols of increasing stringency (decreasing salt concentration and increasing temperature) are needed. Two sequential washing procedures are described below giving medium to high stringency that is sufficient for detection of specific hybridization using genomic DNA probes. They can be adapted to the required stringency by varying the SSC concentrations and the temperature.

Oligonucleotide probes

With oligonucleotide probes (oligoprobes) titration experiments are required to optimize the conditions for which the signal is visible and when it disappears. (Essentially this establishes the T_m for the particular probe.)

Post-hybridization washes

After removing the Gel bond coverslips with a scalpel blade, wash (with agitation) in one of the following sequences of solutions, either (a) or (b) for genomic probes or (c) for oligoprobes:

(a) for genomic probes (high stringency):
   (i) 2 × SSC at room temperature for 10 min;
   (ii) 2 × SSC at 60°C for 20 min;
   (iii) 0.2 × SSC at room temperature for 10 min;
   (iv) 0.2 × SSC at 42°C for 20 min;
   (v) 0.1 × SSC at room temperature for 10 min;
   (vi) 2 × SSC at room temperature for 1–2 min.

(b) for genomic probes (low stringency):
 (i) fresh 4 × SSC at 22°C for 5 min;
 (ii) fresh 4 × SSC at 22°C for 5 min;
 (iii) fresh 4 × SSC at 22°C for 5 min.

(c) for oligoprobes (this is roughly dependent on the T_m of the probe)
 (i) 2 × SSC at room temperature for 5 min;
 (ii) 0.2 × SSC at 42°C for 20 min;
 (iii) 0.1 × SSC at 45°C for 10 min;
 (iv) 0.1 × SSC at 55°C for 10 min.

12.6 Detection of hybridization signal

Many detection techniques are now available for detecting biotinylated and digoxigenin-labelled human and viral DNA probes. For convenience, these are here called one-step, two-step, three-step, and five-step procedures, depending on the number of reagents used and steps involved in the detection protocol. The use of immunogold silver staining for the detection of biotinylated probes is also described. The development reagents given in *Protocol 10* are required for detection of hybridization signals using these protocols.

Protocol 10. Preparation of development reagents

A. *NBT/BCIP development reagent for alkaline phosphatase detection*

Reagents

- Nitroblue tetrazolium (NBT): 75 mg/ml in dimethylsulfoxide (DMSO)
- Bromochloroindoyl phosphate (BCIP): 50 mg/ml in DMSO
- Buffer 1: 0.1 M Tris–HCl pH 9.5, 0.1 M NaCl, 50 mM $MgCl_2$

Method

1. Just before use, mix 3.75 ml of Buffer 1 and 16.5 μl of NBT.
2. Mix gently by inversion.
3. Add 12.5 μl of BCIP.
4. Store this buffer in the dark.

B. *Aminoethylcarbazole (AEC) development reagent for peroxidase detection*[a]

Reagents

- AEC (Sigma)
- DMSO (Merck, UK)
- 20 mM sodium acetate buffer, pH 5.0–5.2
- Hydrogen peroxide, 30% (v/v) (Merck, UK)

Method

1. Dissolve 2 mg of AEC in 1.2 ml of DMSO in a glass tube.
2. Add 10 ml of 20 mM acetate buffer, pH 5.0–5.2.

3. Invert the glass tube and add 1 ml 30% (v/v) hydrogen peroxide. Development occurs over 10–20 min.

[a] Alternatively, use the aminoethylcarbazole substrate kit from Zymed, USA.

12.6.1 Detection of biotinylated probes

Protocols 11-14 describe different detection protocols for biotinylated probes. The choice of detection system depends on the level of sensitivity required.

Protocol 11. One-step procedure for detecting biotinylated probes

Equipment and reagents

- Buffer 2: 0.1 M Tris–HCl pH 7.5, 0.1 M NaCl, 2 mM $MgCl_2$, 0.05% Triton X-100
- Tris-buffered saline (TBS): 50 mM Tris–HCl, 100 mM NaCl, pH 7.2
- Avidin/alkaline phosphatase or avidin peroxidase (Dako, UK), diluted 1/100 in TBT
- Tris-buffered saline/Triton X-100 (TBT): 50 mM Tris–HCl, 100 mM NaCl, pH 7.2, 3% (w/v) BSA, 0.5% Triton X-100 (v/v)
- NBT/BCIP or AEC development reagent (see *Protocol 10*)

Reagents

1. Immerse the slides from *Protocol 9* in TBT (blocking reagent) at 22°C for 2 min.
2. Transfer the slides to an incubation tray (transfer one slide at a time to prevent dehydration).
3. Incubate slides in avidin/alkaline phosphatase or avidin peroxidase, diluted 1/100 in TBT.
4. Remove unbound conjugate by washing for 5 min twice in TBS.
5. Incubate slides in NBT/BCIP or AEC development reagent for 10–30 min and monitor colour development.
6. Terminate the colour development reaction by washing in distilled water for 5 min.

Protocol 12. Two-step procedure for detecting biotinylated probes

Reagents

- Buffer 1 (see *Protocol 10*)
- Buffer 2 (see *Protocol 11*)
- Buffer 2 containing 5% (w/v) BSA
- Avidin DN (Vector), 10 μg/ml in Buffer 2
- Biotinylated alkaline phosphatase (Vector), 10 μg/ml in Buffer 2
- NBT/BCIP development reagent (see *Protocol 10*)

Protocol 12. *Continued*

Method

1. Transfer the slides from *Protocol 9* into Buffer 2 containing 5% (w/v) BSA and incubate at room temperature for at least 30 min.
2. Wipe excess buffer from the slides and transfer them to a slide incubation tray (transfer only a few slides at a time to prevent sections from drying out).
3. Add a few drops of Buffer 2 containing Avidin DN and incubate at room temperature for 10 min.
4. Wash the slides with agitation with two changes of Buffer 2 for 10 min each.
5. Return the slides to the incubation tray and add a few drops of Buffer 2 containing biotinylated alkaline phosphatase to each section. Incubate at room temperature for 10 min.
6. Wash the slides with agitation twice with Buffer 2 for 10 min each.
7. Transfer the slides into Buffer 1 and allow to equilibrate for 30 min.
8. Return the slides to the incubation tray and cover the sections with NBT/BCIP development reagent. Monitor the development of the colour after 5 min and then continuously until development looks complete.
9. Terminate the reaction by immersion of slides in PBS for 5 min or in distilled water.

Protocol 13. Three-step procedure for detecting biotinylated probes

Reagents

- TBT (see *Protocol 11*)
- TBS (see *Protocol 11*)
- Monoclonal mouse anti-biotin antibody (Dako, UK) diluted 1 in 50 in TBT
- Biotinylated rabbit anti-mouse $F(ab')_2$ fragment (Dako, UK) diluted 1 in 20 in TBT
- Either avidin alkaline phosphatase (Dako) diluted 1 in 50 in TBT, or streptavidin peroxidase (Dako, UK) diluted 1 in 100 in TBT containing 5% non-fat milk
- Development reagent: either NBT/BCIP or AEC (see *Protocol 10*)

Method

1. Immerse the slides from *Protocol 9* in TBT at 22°C for 2 min.
2. Transfer the slides to an incubation tray and incubate in monoclonal mouse anti-biotin diluted 1 in 50 in TBT.
3. Wash the slides twice in TBS for 10 min each.
4. Incubate slides in biotinylated rabbit anti-mouse $F(ab')_2$ fragment diluted 1 in 20 in TBT.

5. Wash twice in TBS for 5 min.
6. Incubate the slides in either avidin alkaline phosphatase or streptavidin peroxidase.
7. Wash in TBS for 5 min.
8. Incubate the slides in either NBT/BCIP development reagent or AEC for 10–30 min, as appropriate.
9. Terminate the colour development reaction by washing in distilled water for 5 min.

Protocol 14. Detection of probe with gold-labelled goat anti-biotin

Equipment and reagents

- Dark room (with S902 or F904 safelight)
- Lugol's iodine (Sigma)
- 2.5% (w/v) aqueous sodium thiosulfate
- TBS (see *Protocol 11*)
- TBS containing 0.8% (w/v) BSA, 0.1% (w/v) gelatin, 5%(v/v) normal swine serum, and 2 mM sodium azide
- Citrate buffer prepared by dissolving 2–3.5 g of trisodium citrate dihydrate and 25.5 g citric acid monohydrate in 100 ml of distilled water; adjust to pH 3.5
- 1 nm gold-labelled goat anti-biotin antibody (1:10 Auroprobe in TBS containing 0.1% (w/v) gelatin, 0.8% (w/v) BSA, 1% (v/v) normal swine serum (Sigma), and 2 nM sodium azide)
- Silver developing solution prepared in a dark room by mixing 7.5 ml of gum acacia (500 g/l) diluted to 60 ml with water, 10 ml of citrate buffer, pH 3.5, 15 ml of hydroquinone (0.85 g/15ml), and 15 ml silver lactate (0.11 g/15 ml)

Method

1. Transfer the slides from *Protocol 9* into Lugol's iodine for 2 min.
2. Rinse with TBS, decolorize in 2.5% (w/v) aqueous sodium thiosulfate, and wash in TBS twice for 5 min.
3. Immerse the slides in TBS containing 0.8% (w/v) BSA, 0.1% (w/v) gelatin, 5% (v/v) normal swine serum, and 2 mM sodium azide for 20 min.
4. Remove excess buffer and transfer slides from the incubation tray. Apply 1 nm-gold-labelled goat antibody solution and incubate for 2 h.
5. Wash the sections in TBS with agitation twice for 5 min.
6. Immerse the slides in silver developing solution in the dark, until the sections appear optimally developed, when viewed by light microscopy (3–10 min).
7. Wash in running tap water for 5 min.
8. Fix in 2.5% (w/v) aqueous sodium thiosulfate for 3 min.
9. Wash in running tap water for 1 min, counter-stain as required, dehydrate, clear, and mount with synthetic resin.

12.6.2 Detection of digoxigenin-labelled probes

Protocols 15–17 describe techniques for detecting digoxigenin-labelled probes. The choice of detection system depends on the level of sensitivity required—the greater the number of steps, the more sensitive the technique.

Protocol 15. One-step procedure for detecting digoxigenin-labelled probes

Reagents

- TBT (see *Protocol 11*)
- TBS (see *Protocol 11*)
- NBT/BCIP developing reagent (see *Protocol 10*)
- Alkaline phosphatase-conjugated anti-digoxigenin antibody (Boehringer, Germany) diluted 1 in 600 in TBT

Method

1. Immerse the slides from *Protocol 9* in TBT at 22°C for 10 min.
2. Transfer the slides into an incubation tray.
3. Incubate the sections in alkaline phosphatase-conjugated anti-digoxigenin antibody for 30 min at room temperature.
4. Wash in TBS for 5 min, twice.
5. Develop the signal using NBT/BCIP development reagent for 10–30 min and monitor colour development.
6. Terminate the colour development reaction by washing in distilled water for 5 min.

Protocol 16. Three-step procedure for detecting digoxigenin-labelled probes

Reagents

- TBT (see *Protocol 11*)
- TBS (see *Protocol 11*)
- Monoclonal anti-digoxin antibody (Sigma, UK) diluted 1 in 10 000 in TBT
- Biotinylated rabbit anti-mouse $F(ab')_2$ fragment (Dako) diluted 1 in 200 in TBT
- Avidin alkaline phosphatase (Dako), diluted 1 in 50 in TBT, or avidin/peroxidase (Dako) diluted 1 in 75 in TBT containing 5% (w/v) skimmed milk
- NBT/BCIP or AEC development reagent (see *Protocol 10*)

Method

1. Immerse the slides from *Protocol 9* in TBT (blocking reagent) at 22°C for 10 min.
2. Transfer the slides to an incubation tray.
3. Incubate the slides in monoclonal anti-digoxin antibody for 30 min at room temperature.

4. Wash in TBS twice for 5 min.
5. Incubate in biotinylated rabbit anti-mouse $F(ab')_2$ fragment for 30 min at room temperature.
6. Wash in TBS twice for 5 min.
7. Incubate the slides in avidin alkaline phosphatase or avidin/peroxidase for 30 min at room temperature.
8. Wash in TBS for 5 min.
9. Incubate the slides in NBT/BCIP or AEC development reagent as appropriate.
10. Terminate the colour development reaction by washing in distilled water for 5 min.

Protocol 17. Five-step procedure for detecting digoxigenin-labelled probes

Reagents

- TBT (see *Protocol 11*)
- TBS (see *Protocol 11*)
- Monoclonal anti-digoxin antibody (see *Protocol 16*)
- Biotinylated rabbit anti-mouse $F(ab')_2$ fragment (see *Protocol 16*)
- Monoclonal anti-biotin antibody (Dako, UK) diluted 1 in 50 in TBT
- Avidin peroxidase diluted 1 in 75 in TBT containing 5% skimmed milk
- AEC development reagent (see *Protocol 10*)

Method

1. Place in TBT and incubate with monoclonal anti-digoxin- and biotin-labelled rabbit anti-mouse antibody as described in *Protocol 16*, steps 1–6.
2. Incubate the slides in monoclonal anti-biotin antibody.
3. Wash slides in TBS for 5 min.
4. Incubate the slides in biotinylated rabbit anti mouse $F(ab')_2$ fragment (Dako, UK) for 4 min.
5. Wash in TBS for 5 min.
6. Incubate the slides in avidin peroxidase diluted to 1 in 75 in TBT containing 5% skimmed milk.
7. Incubate the slides in the AEC development reagent for 10–30 min and monitor colour development.
8. Terminate the colour development reaction by washing in distilled water for 5 min.

12.6.3 Simultaneous detection of biotin- and digoxigenin-labelled probes

Two different nucleic acids can be detected simultaneously in one tissue section using combined application of biotin- and digoxigenin-labelled probes to their respective targets in the sample, as detailed in *Protocol 18*.

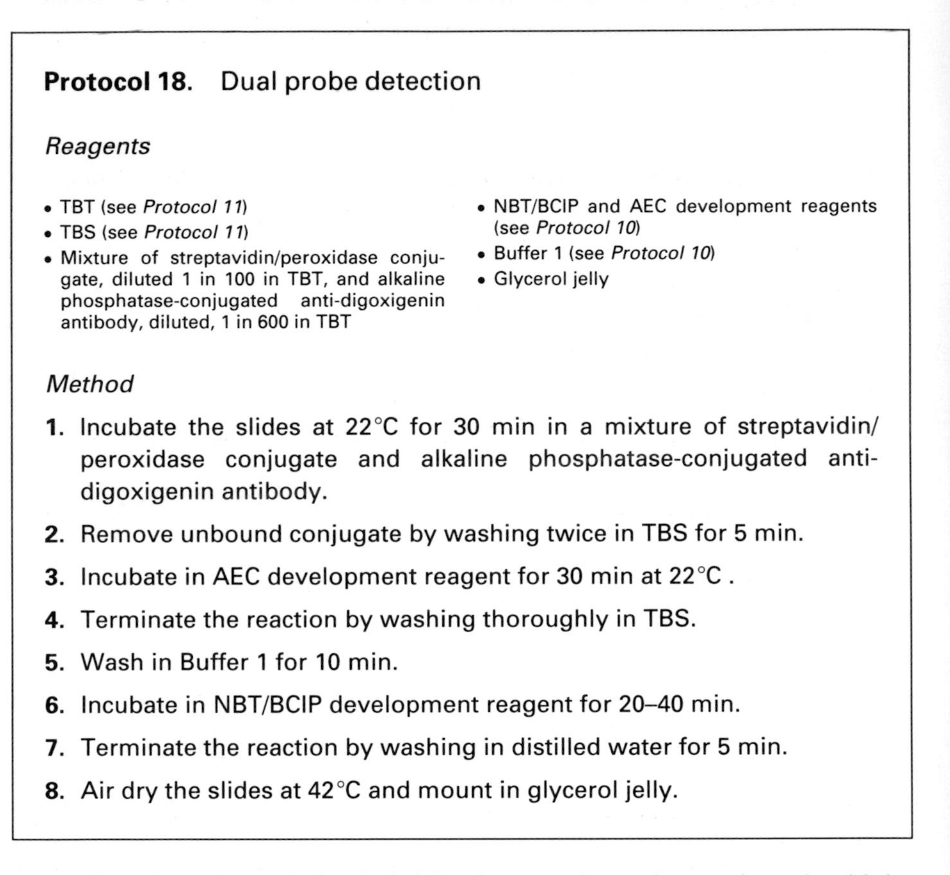

Protocol 18. Dual probe detection

Reagents

- TBT (see *Protocol 11*)
- TBS (see *Protocol 11*)
- Mixture of streptavidin/peroxidase conjugate, diluted 1 in 100 in TBT, and alkaline phosphatase-conjugated anti-digoxigenin antibody, diluted, 1 in 600 in TBT
- NBT/BCIP and AEC development reagents (see *Protocol 10*)
- Buffer 1 (see *Protocol 10*)
- Glycerol jelly

Method

1. Incubate the slides at 22°C for 30 min in a mixture of streptavidin/peroxidase conjugate and alkaline phosphatase-conjugated anti-digoxigenin antibody.
2. Remove unbound conjugate by washing twice in TBS for 5 min.
3. Incubate in AEC development reagent for 30 min at 22°C .
4. Terminate the reaction by washing thoroughly in TBS.
5. Wash in Buffer 1 for 10 min.
6. Incubate in NBT/BCIP development reagent for 20–40 min.
7. Terminate the reaction by washing in distilled water for 5 min.
8. Air dry the slides at 42°C and mount in glycerol jelly.

Following detection of the hybridization product, the section should be counter-stained with an aqueous stain. In the case of alkaline phosphatase (NBT/BCIP) detection, counter-staining with 2% (w/v) Methyl Green and mounting with glycerol jelly should be performed. All slides detected using AEC should be counter-stained progressively with haematoxylin for approximately 10–15 sec and then mounted in glycerol jelly.

12.6.4 Detecting fluorescent probes

Probes labelled with fluorescein isothiocyanate (FITC) can be detected either directly or by conversion to a colorimetric end-product using immunodetection methods such as that detailed in *Protocol 19*.

Protocol 19. Detection of FITC-labelled probes

Reagents

- TBT (see *Protocol 11*)
- TBS (see *Protocol 11*)
- Normal rabbit serum (Sigma) diluted 1 in 50 in TBT
- Buffer 1 (see *Protocol 10*)
- Rabbit alkaline phosphatase-conjugated anti-FITC antibody (Nova Castra) diluted 1 in 100 or 1 in 200 in TBT
- NBT/BCIP development reagent (see *Protocol 10*)

Method

1. Place slides in an incubating tray, cover the sections with 100 ml of normal rabbit serum (Sigma) diluted 1 in 50 in TBT, and incubate for 10 min at room temperature.
2. Remove the slide carefully and add alkaline phosphatase-conjugated rabbit anti-FITC antibody. Incubate for 30 min.
3. Wash the slides in TBS twice for 3 min.
4. Wash the slides in Buffer 1 for 5–10 min.
5. Incubate the slides in NBT/BCIP development reagent for 10–15 min as appropriate.
6. Terminate the colour development by washing in distilled water for 5 min.

References

1. Herrington, C.S., de Angelis, M., Evans, M.F., Troncone, G., and McGee, J. O'D. (1992). *J. Clin. Pathol.*, **45**, 385.
2. Haase, A.T., Retzel, E.F., and Staskus, K.A. (1990). *Proc. Natl Acad. Sci. USA*, **87**, 4971.
3. Nuovo., G.J. (1992). *PCR in situ hybridisation. Protocols and applications*. Raven Press, New York.
4. Nuovo, G.J., Gallery, F., MacConnell, P., Becker, P., and Bloch, W. (1991). *Am. J. Pathol.*, **139**, 1239.
5. O'Leary, J.J., Browne, G., Johnson, M.I., Landers, R.J., Crowley, M., Healy, I. *et al.* (1994). *J. Clin. Pathol.*, **47**, 933.
6. Embretson, J., Zupancic, M., Beneke, J., Till, M., Wolinsky, S., Ribas, J.L., Burke, A., and Haase, A.T. (1993). *Proc. Natl Acad. Sci. USA.*, **90**, 357.
7. Gosden, J., Hanratty, D., Starling, J., Fantes, J., Mitchell, A., and Porteous, D. (1991). *Cytogenet. Cell Genet.*, **5aa7**, 100.
8. Spann, W., Pachmann, K., Zabnienska, H., Pielmeier, A., and Emmerich, B. (1991). *Infection*, **19**, 242.
9. Komminoth, P., Long, A.A., Ray, R., and Wolfe, H.J. (1992). *Diagn. Mol. Pathol.*, **1**, 85.
10. Hopwood, D. (1985). *Histochem. J.*, **17**, 389.

11. Bagasra, O., Hauptman, S.P., Lischner, H.W., Sachs, M., and Pomerantz, R.J. (1992). *New Engl. J. Med.*, **326**, 1385.
12. Chiu, K., Cohen, S.H., Morris, D.W., and Jordan, G.W. (1992). *J. Histochem. Cytochem.*, **40**, 333.
13. Embleton, M.J., Gorochov, G., Jones, P.T., and Winter, G. (1992). *Nucleic Acids Res.*, **20**, 3831.
14. Staskus, K.A., Couch, L., Bitterman, P., Retzel, E.F., Zupancic, M., List, J., and Haase, A.T. (1991). *Microbiol. Pathol.*, **11**, 67.
15. Long, A.A., Komminoth, P., and Wolfe, H.F. (1993). *Histochemistry*, **99**, 151.
16. Ray, R., Komminoth, P., Macado, M., and Wolfe, H.J. (1991). *Mod. Pathol*, **4**, 124A.
17. Yap, E.P.H. and McGee, J.O'D. (1991). *Nucleic Acids Res.*, **19**, 4294.
18. Hopman, A.H.N., Speel, E.J.M., Voorter, C.E.M., and Ramaekers, R.C.S. (1995). Probe labelling methods. In *Non-isotopic methods in molecular biology: a practical approach* (ed. E.R. Levy and C.S. Herrington), pp. 1–24. IRL Press, Oxford.

5

Reverse transcriptase *in situ* PCR for RNA detection

M. JIM EMBLETON

1. Introduction

Most published applications of *in situ* PCR for amplification of specific nucleic acid sequences in cells or tissues have been for the detection of genes or viruses which are present in low copy numbers or in a minority of the cell population. Detection may be either indirect, for example by hybridizing a labelled oligonucleotide probe to the amplified DNA, or direct, where the product is amplified with labelled oligonucleotide primers. A variety of labels for oligonucleotides have been used in different situations, including fluorescent dyes for direct microscopic visualization (1,2), and ligands such as biotin or digoxigenin (3–5), which can be used for visualization by other means involving enzyme–substrate systems or binding of labelled antibodies. An application of *in situ* PCR which so far has been less well exploited is the production of cDNA for cloning. For most cloning applications amplification from nucleic acid templates extracted from cells or tissues is perfectly adequate, and is technically easier to achieve than *in situ* PCR. However, there are some situations in which it is desirable to co-amplify two separate gene sequences within the same cell, and to maintain them in some form of spatial or physical linkage. The options for this are to carry out PCR on single cells isolated from each other (e.g. in single wells of a microtitre plate), or intracellular PCR on a population of cells. Where the target genes comprise a very large heterogeneous pool, it is impractical to establish huge numbers of single-cell PCRs. A method was therefore devised to perform the PCR *in situ* on a population of fixed and permeabilized cells which maintained their integrity throughout the process, and to arrange for the amplified sequences to be linked physically within the cells so that they could be recovered in the combinations originally expressed by the cells (1). The specific purpose of this technique was the rescue and linkage of immunoglobulin variable region (V) genes from a population of B lymphocytes containing a multitude of different rearranged immunoglobulin genes, in order to prepare libraries encoding recombinant antibody fragments representative of the original repertoire. In the published studies, after two separate PCRs, the amplified linked cDNA was present in the PCR supernatant in sufficient quantities to allow cloning and confirmation of the fidelity of chain pairing, using a model system (1). Importantly, it was also found that about 30% of the product was retained in the cells. In order to provide positive demonstration of intracellular linkage of the V_H and V_L genes, the product was therefore further amplified using fluorescently labelled primers, in order to allow direct visualization by confocal fluorescence microscopy. Although this step was carried out initially for experimental verification of the linkage mechanism, it forms the basis of an *in situ* method for RNA detection.

Examples described in this chapter refer to detection of PCR-linked immunoglobulin V genes within hybridoma cells, and the *bcr–abl* fusion gene (characteristic of chronic myeloid leukaemia cells) in the human myeloid leukaemia cell line K562. However, the basic method is equally applicable to other genes or mRNA species in other cell types. Also, labelling methods other than fluorescence are possible. Oligonucleotides labelled with biotin or digoxigenin can be incorporated into DNA sequences by PCR, and the products detected using enzymes conjugated to streptavidin or anti-digoxigenin antibody, respectively. Alternatively, the product can be hybridized with a labelled specific oligonucleotide probe. Because of the originally intended purpose of the technique, it is aimed particularly at PCR of cell suspensions in tubes rather than tissue sections. However, it could be modified for amplification and labelling of mRNA expressed in cells in sections or smears mounted on microscope slides, by using a flat-bed thermal cycler designed for slides, with humidification or sealing devices to prevent evaporation of the reagent mix from the sample (see Chapter 7).

The basic published method (1) involved preparation of formalin-fixed and permeabilized cells, and carrying out first-strand cDNA synthesis (reverse transcription, RT) in the cells. This was followed by a series of three PCRs, the first for amplification and/or linkage of specific cDNAs, the second a nested PCR to append tags complementary to 'universal' fluorescent primers, and the third to amplify the second PCR product with the fluorescent primers. As pointed out above, the protocols described were initially intended to produce soluble PCR products for cloning, and were incidentally found to be suitable also for RNA detection. Other workers have partially optimized fixation and permeabilization techniques to minimize leakage of product from the cells, in order to improve sensitivity and specificity of *in situ* detection (6). The methodology can perhaps, therefore, be 'tuned' for either detection or cloning.

2. Preparation of cells for RT-PCR (see *Protocol 1*)

The following protocols are designed for cell suspensions, and are suitable for cells harvested from culture, isolated from blood or other body fluids, or prepared by mechanical or enzymatic dissociation of tissues. Because of the variable nature of these sources, and the fact that methods of preparing cells from them are already standard in most laboratories equipped for RT-PCR, protocols for their isolation are not given here; the availability of a cell suspension is therefore assumed. The RT and PCR reactions occur in the cell cytoplasm, so it is important that the cells are intact and of high viability from the outset, and that they are in the form of a single-cell suspension. If substantial clumps are present throughout the procedure, PCR products leaking from positive cells can diffuse into the cytoplasm of adjoining negative cells, giving a high risk of false positives. Another requirement is that adequate numbers of cells are available, because losses can occur during each washing step. A suggested minimum is 10^6 cells, but substantially more would be easier to handle. Also, it is convenient to prepare larger batches of fixed and permeabilized cells because they can be stored at –70°C for future use. If fewer than 10^6 cells are available they could be used in the form of smears on slides or coverslips, with adaptation of the protocols. Before fixation, all cells are fragile and should be handled gently during the washes. After fixation they are much more resistant to mechanical damage and can be handled quite vigorously. For example, clumps caused by cross-linking during aldehyde fixation may be dispersed by syringing through a fine needle, without harming the cells. Microscopically the cells are more refractile and may appear less spherical after fixation and permeabilization than in their natural state, but the change in appearance does not indicate significant damage.

Protocol 1. Preparation of fixed and permeabilized cell suspensions

Equipment and reagents

- Bench-top centrifuge
- Centrifuge tubes (10–50 ml)
- Microcentrifuge
- Haemocytometer (counting chamber) and microscope
- Rotator in a 4°C cold-room, if available
- Set of pipettes and tips
- 1 ml syringes, 21 and 26 gauge needles
- Ice bucket
- 2 ml Eppendorf 'Safe-Lock' tubes
- 10% formal saline: 0.15 M NaCl containing 10% (v/v) formalin solution (formalin = 40% formaldehyde), equivalent to 4% formaldehyde in final concentration
- 0.5% Nonidet P-40 (NP-40) in distilled water
- Phosphate-buffered saline (PBS): 10 mM phosphate buffer, pH 7.2–7.4, with 0.15 M NaCl
- PBS with 0.1 M glycine (PBS-G)
- Cell suspension
- –70°C freezer

Method

1. Wash the cells three times in PBS, centrifuging in a bench centrifuge at 100*g* for 5 min between washes.
2. Count the cells in a haemocytometer and aliquot up to 10^7 cells into a 2 ml Eppendorf tube. Spin at 16 000*g* in a microcentrifuge for 2.5 min.
3. Remove the supernatant and suspend the cell pellet in 1 ml ice-cold

Protocol 1. *Continued*

10% formal saline, using a syringe and 21 gauge needle. To minimize cell losses, retain the syringe and needle for re-use. Incubate the tube at 4°C on a rotator, or on ice with frequent shaking for 1 h.

4. Spin the tube in a microcentrifuge at 16 000*g* for 2.5 min and remove the supernatant. Suspend the cells in ice-cold PBS-G, using the syringe and needle from step 3. Repeat the centrifugation and resuspension for a total of three washes.
5. Resuspend the cells in 1 ml ice-cold 0.5% NP-40 using the same syringe and needle, and incubate for 1 h on a rotator at 4°C or on ice with frequent shaking.
6. Spin down and wash three times as in step 4.
7. Finally resuspend the cells in ice-cold PBS-G using the 1 ml syringe, but changing to a 26 gauge needle. Pass the cells through the needle five or six times.
8. Count the cells on a haemocytometer, checking that they are free of microscopic clumps. If clumps are present, re-syringe the cells until they are dispersed. Adjust the concentration to 2×10^6 cells/ml. Store in aliquots at –70°C, or proceed directly to *Protocol 2*.

3. Reverse transcriptase reaction (first-strand cDNA synthesis)

The reverse transcriptase reaction (see *Protocol 2*) may be performed on freshly prepared or freshly thawed fixed and permeabilized cells. It can be performed equally successfully using avian myeloblastosis virus (AMV) or Moloney murine leukaemia virus (MMLV) reverse transcriptases. However, AMV reverse transcriptase requires an incubation temperature of 42°C and MMLV reverse transcriptase requires incubation at 37°C. Manufacturers normally supply a concentrated buffer for use with their transcriptase, but the protocol also describes a 10 × first-strand buffer which has been found to work satisfactorily with both AMV and MMLV reverse transcriptase. If carried out on a thermal cycler, the reaction can be performed in 0.5 ml tubes, but if carried out in a heat block with larger holes or a water bath, it is better to use 2 ml Eppendorf 'Safe-Lock' tubes which provide a higher surface area at the bottom. Cells will settle to a loose pellet under gravity during the reaction. The RT reaction can be primed using random oligo(dT) primers or custom-designed forward (anti-sense) primers 3′ to the gene of interest. Up to five primers for different RNA sequences have been used successfully in the same reaction. Diethylpyrocarbonate-treated water may be used instead of distilled water if desired, but has not been found to be necessary.

Protocol 2. Reverse transcription *in situ*

Equipment and reagents

- Microcentrifuge
- Adjustable pipettes and sterile tips
- Water bath(s) or heat block(s) at 37°C or 42°C, and 65°C; alternatively, a programmable thermal cycler can be used
- 0.5 ml Eppendorf or Sarstedt tubes and 2 ml Eppendorf 'Safe-Lock' tubes
- PBS-G (see *Protocol 1*)
- Promega RNasin (other RNase inhibitors may work, but have not been tested)
- Promega MMLV reverse transcriptase, or Promega AMV reverse transcriptase, or HT Biotechnology AMV reverse transcriptase (other brands should work, but have not been tested); buffer concentrates are provided by these manufacturers
- Sterile double-distilled water
- Forward (anti-sense or 3') oligonucleotide primers, diluted in water to 10 pmol/ml
- dNTP mix containing 5 mM each of dATP, dCTP, dGTP, and dTTP in distilled water; the overall dNTP concentration is thus 20 mM (dNTPs from Promega, Pharmacia, and Boehringer Mannheim have all been used successfully)
- 10 × first-strand buffer, if not using a manufacturer's buffer: 1.4 M KCl, 0.5 M Tris–HCl pH 8.1 at 42°C, 80 mM $MgCl_2$
- Dithiothreitol (DTT) 100 mM solution, if not using a manufacturer's first-strand buffer (the buffer concentrates from the specified manufacturers already contain DTT); 1 M stock solutions may be frozen in 0.01 M sodium acetate and diluted in distilled water
- Fixed and permeabilized cells from *Protocol 1*
- Ice bucket

Method

1. Prepare a 'first-strand' mix in a 0.5 ml or 2 ml tube as follows, and place on ice:
   - 2 μl of each primer
   - 5 μl of dNTP mix
   - 2 μl of RNAsin (80 units)
   - 5 μl of 10 × first-strand buffer or manufacturer's 10 × buffer (HT); or 10 μl of manufacturer's 5 × buffer (Promega)
   - (*only* if using own 10 × first-strand buffer) 5 μl of DTT
2. Spin down an aliquot of 1×10^6–2×10^6 fixed and permeabilized cells in a 0.5 ml tube for 2.5 min at 16 000*g* in a microcentrifuge. Remove the supernatant and suspend the cells in 200 μl of distilled water. Spin down again and remove the supernatant. Resuspend the cells in sufficient water to make up a total volume of 48 μl when added to the first-strand mix (step 3).[a]
3. Add the cells to the first-strand mix to make a total volume of 48 μl, and mix by gentle pipetting.
4. Incubate the cells at 65°C for 5 min, and allow them to cool gradually to below 30°C. Place the tube on ice for a further 5 min.
5. Add 2 μl (40 units) of reverse transcriptase to the cells and mix by gentle pipetting. The total volume should now be 50 μl.
6. Incubate for 1 h at 37°C if MMLV reverse transcriptase is used, or 1 h at 42°C if AMV reverse transcriptase is used. Gently shake the cells to

Protocol 2. *Continued*

suspend them at 10 min intervals (easier with 2 ml tubes), taking care not to splash cells and reagents up the sides of the tube.

7. Transfer the whole cell suspension to a 0.5 ml tube and spin the cells down in a microcentrifuge at 16 000*g* for 2.5 min. The cell pellet should be clearly visible at the bottom of the tube, and will be displaced towards the side which was outermost in the microcentrifuge. It is important for the next step that its orientation is clear, and this can be aided by always spinning the tubes in the same position, e.g. with the hinge of the cap facing outwards.
8. Remove all of the supernatant using a pipette with a fine tip, as follows. Set the pipette to 200 μl and place the tip at the bottom of the tube, next to the cell pellet but not directly on it. Withdraw the supernatant slowly and continuously, without disturbing the cell pellet. When withdrawing the last few microlitres of supernatant, slightly invert the tube and draw the pipette tip along the side of the tube opposite to the cell pellet, from the bottom to the top, while continually applying suction. This procedure should leave the cell pellet and the walls of the tube free of residual liquid, rendering more than a single wash unnecessary. It is advisable to practise the technique with spare fixed cells before attempting an experiment for the first time. Washing is essential, but must be kept to a minimum to avoid undue loss of cells in the pipette tips; a stringent wash, leaving no visible supernatant, is thus very important. The supernatant can be discarded. It may contain a little cDNA, but most of the cDNA is not released unless the cells are boiled.
9. Wash the cells by suspending them in 200 μl of PBS-G and spinning again at 16 000*g* for 2.5 min. Withdraw the supernatant as described in step 8. Finally resupend the cells in 100 μl PBS-G. They can be used immediately in PCR, or frozen in 10 μl aliquots at –70°C.

[a] The actual volume of water for resuspending the cells must be calculated to take into account variations in the volume of buffer concentrate, whether DTT is added, and the number of primers added to the first-strand mix. For example, if three primers are used (total 6 μl) and Promega 5 × buffer (10 μl) with no added DTT, the volume of the first-strand mix will be 23 μl. The cells must therefore be suspended in 25 μl of water before being added to the first-strand mix to give a total volume of 48 μl.

4. PCR with fluorescent primers

The protocol which will be described in detail (see *Protocol 3*) involves three consecutive PCRs on fixed and permeabilized cells containing cDNA reverse transcribed from an mRNA template (*Figure 1*). The first PCR amplifies the cDNA sequence of interest, and the second is a nested PCR which appends a 5′ tag for subsequent recognition by 'universal' fluorescently labelled primer(s), in a third PCR.

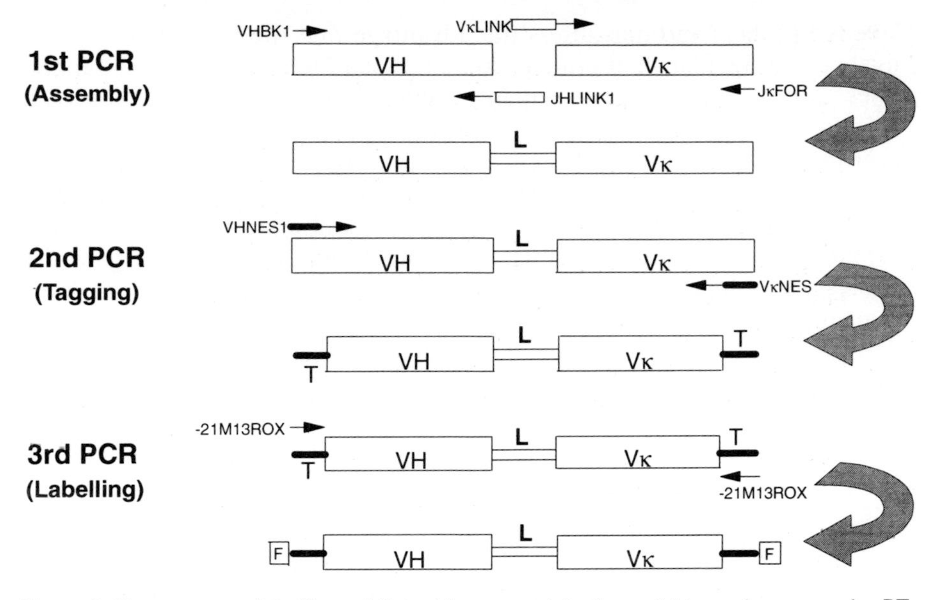

Figure 1. Fluorescence labelling of linked immunoglobulin variable region genes by RT-PCR. The figure portrays the three PCR steps involved in fluorescence labelling of cDNA corresponding to the linked immunoglobulin V regions of hybridoma NQ10/12.5. The first PCR is preceded by an RT reaction using primer J_HFOR for the heavy chain and primer J_κFOR for the kappa chain (not shown). For each separate PCR, the top diagram indicates the priming positions of the oligonucleotides, and the one immediately below portrays the cDNA product of that PCR. Each PCR primer is shown as a short arrow, with its designation in text beside it. All primers are identified in *Tables 1* and *2*, and the sequences listed in *Table 2*. The linker sequence in first PCR primers is shown as an open rectangle, and the tagging sequence in the second PCR primers is shown as a closed rectangle. Abbreviations are as follows: V_H, heavy chain V gene DNA; V_κ, kappa chain V gene DNA; L, the linker sequence produced by annealing of the complementary tags on the linker primers; T, the tag sequence which provides the template for the third PCR primer; F (in an open square), the fluorescent label appended by the third PCR primer

The method was designed in this way because its original purpose was to link two separate cDNA sequences (immunoglobulin V_H and V_L genes) in the first PCR, and to amplify the linked product in preference to single domains in a second PCR, at the same time as adding the tags. Nesting of the second PCR primers was necessary to achieve efficient and highly specific amplification of the linked product. A similar protocol was also adopted for detection of the *bcr–abl* fusion gene in K562 cells. For a single gene, the tags could be added in one PCR to reduce the whole procedure, including subsequent amplification with the fluorescent primer(s), to a two-PCR protocol. However, the specificity of a two-PCR protocol has not been verified, and an intermediate nested PCR undoubtedly increases specificity. The use of labelled primers directly in a one-PCR protocol has been reported to lead to poor specificity (5), and is definitely not recommended. The use of 'universal' labelled primers and tags offers the advantage that the same batch of (relatively expensive) labelled primers could be used for a wide variety of applications. Also, if only one gene is being amplified, the same tag can be used for both forward and backward primers, which enables the use of a single fluorescent primer in the final PCR and results in a strong fluorescence signal. For detection of two independent genes in the same cell population, one tag and fluorescent primer can be used for one gene, and a different tag and fluorescent primer with a different colour fluorochrome for the other gene. The fluorescent primers used in our studies (1) were commercially available cycle-sequencing primers, but tags and matching fluorescent primers custom-designed by the user would be equally suitable and probably less expensive. The sequence of the tag region is irrelevant, and it could be advantageous to design it to avoid any possible mis-priming in the gene(s) of interest, or to avoid possible conflicts with patents or commercial interests. Kits or reagents for preparing fluorescently labelled oligonucleotides or labelling existing oligonucleotides, plus instructions for use, are obtainable from Promega, Boehringer Mannheim, Sigma, and Perkin–Elmer.

The use of fluorescence labelling requires specialized equipment to detect cells containing the target sequence. At the minimum level, this means a conventional fluorescence microscope with filters suitable for either fluorescein or rhodamine, or both. The facility for two-colour fluorescence allows detection of two different mRNA sequences, either in different cells or in the same cells, which could be an advantage for a number of applications. A confocal fluorescence microscope offers a greater degree of sophistication by producing images in a single plane, which enables much clearer resolution of the location of fluorescence within the cells and greater sensitivity. In addition to microscopy, fluorescently labelled cells have the great advantage that they can be detected by flow cytometry. This allows accurate enumeration of small numbers of positive cells in a large population, and more importantly the sorting of positive cells into a purified sub-population. Although the sorted cells are not functional, flow cytometry offers the possibility of rescuing genes specifically from a selected population.

The PCR is carried out in tubes, and during the reaction the cells settle into a loose pellet at the bottom of the tube. It is not necessary to resuspend the cells during the reaction, but if desired they could be suspended by gentle pipetting at some point, e.g. during a pause introduced during a denaturation step. However, this could lead to undue loss of cells in the pipette tip if repeated often, and is not recommended. Leakage of PCR products from one cell to another can be a serious problem with cells which are strongly adherent to each other (i.e. cells in clumps), but has not proved to be a problem in the case of cells which have settled loosely together under gravity. The correct washing procedure carried out between PCRs is sufficient to remove any such contamination if it occurs, as testified by the demonstrated fidelity of the technique in published data (1), and in the data summarized in *Table 1*.

Table 1. Specific *in situ* amplification and fluorescence labelling of cDNAs in cell suspensions

Cell line[a] labelling	First PCR[b] primers	Second PCR primers	Third PCR primer	Fluorescence
NQ10/12.5	V_HBK1 + J_HLINK1 + V_κLINK + J_κFOR[c]	V_HNES1 + V_κNES[c]	–21M13ROX	+
			M13 reverse ROX	–
NQ10/12.5	V_HBK2 + J_HLINK2 + V_λLINK + J_λFOR[d]	V_HNES2 + V_λNES[d]	–21M13ROX	–
			M13 reverse ROX	–
B1-8	V_HBK1 + J_HLINK1 + V_κLINK + J_κFOR	V_HNES1 + V_κNES	–12M13 ROX	–
			M13 reverse ROX	–
B1-8	V_HBK2 + J_HLINK2 + V_λLINK + J_λFOR	V_HNES2 + V_λNES	–21M13ROX	–
			M13 reverse ROX	+
K562	ABLC + BCRA[e]	–21BCRB + –21ABL2[e]	–21M13ROX	+
			M13 reverse ROX	–

[a]NQ10/12.5 is a hybridoma cell line secreting a kappa light chain (7); B1-8 is a hybridoma line secreting a lambda light chain (8); K562 is a human myeloid leukaemia cell line characterized by a *bcr–abl* fusion gene resulting from a translocation between the *abl* gene on chromosome 9 and the *bcr* gene on chromosome 22 (9).
[b]Primer sequences are listed in *Table 2*. First PCR 'LINK' primers achieve linkage between heavy and light chain V genes, second PCR primers are nested with respect to first PCR 'BK' and 'FOR' primers, and third PCR primers have a red fluorescent label.
[c]Primer set specific for NQ10/12.5 V genes.
[d]Primer set specific for B1-8 V genes.
[e]Primers specific for the *bcr–abl* fusion gene.

Although the recovery of cells is the most important aim in each PCR, a useful check on the size of the amplified product can be made by running an agarose gel of the DNA present in the supernatant. The isolated DNA could also be cloned to allow additional confirmation of its identity, for example by colony hybridization with a labelled oligonucleotide probe. A separate protocol (*Protocol 4*) describes how to run an agarose gel.

Protocol 3. *In situ* PCR with fluorescence labelled primers

Equipment and reagents

- Microcentrifuge
- Adjustable manual pipettes and sterile tips (or, preferably, aerosol-resistant filter tips)
- Thermal cycler, preferably one that can operate without an oil overlay in the PCR tubes (e.g. a hot lid or hot air machine)
- 0.5 ml Eppendorf or Sarstedt tubes
- Sterile distilled water
- Mineral oil, if an oil-free thermal cycler is not available
- PBS-G (see *Protocol 1*)
- dNTP mix, 5 mM each dNTP (see *Protocol 2*)
- Fixed and permeabilized cells containing cDNA template from *Protocol 2*
- Oligonucleotide primers at 10 pmol/μl (see *Table 2* for specific examples and *Figure 1* for schematic diagram)
- *Taq* DNA polymerase: most brands work well in this technique, but not all are licensed for PCR; Perkin–Elmer and Boehringer Mannheim *Taq* are examples of suitable *Taq* polymerases, and proof-reading polymerases or mixed formulations are also good (e.g. Boehringer-Mannheim Extend Hi-Fidelity). Buffer concentrates are provided by the manufacturers.
- Apparatus for examining fluorescent cells (fluorescence microscope, confocal microscope, or flow cytometer)

Method

1. Prepare a first PCR mix in a 0.5 ml tube as follows:
   - back (5′ or sense) primer, 2.5 μl
   - forward (3′ or anti-sense) primer, 2.5 μl
   - dNTP mix, 2.0 μl
   - *Taq* DNA polymerase (2.5 units), 0.5 μl
   - 10 × *Taq* polymerase buffer, 5.0 μl
   - distilled water, 27.5 μl
   - cells (5×10^4–1×10^5) in PBS-G[a], 10.0 μl

   (total volume of reaction mix = 50 μl)
2. Mix the cell suspension well by gentle pipetting. If using a hot-lid or hot-air thermal cycler, do not use oil; the avoidance of oil simplifies the washing procedure between each PCR. However, if a conventional block cycler is used, add a drop of mineral oil to each tube. Close the cap of the tube very securely.
3. Subject the tubes to 30 PCR cycles as follows:
   - denaturation: 95°C for 1 min
   - annealing: 5°C below the T_m of the primers[b] for 1 min
   - extension: 72°C for 2 min
4. If the PCR was performed without oil, spin the cells down in a micro-

centrifuge at 16000*g* for 2.5 min. Set a pipette fitted with a fine tip to 200 μl and remove all of the supernatant from the cell pellet, using the procedure described in *Protocol 2*, step 8. If intending to check the size of the PCR product by running an agarose gel, keep the supernatant in a new tube.

5. If mineral oil was used in PCR tubes, the cells must be removed from the tube without any oil contamination. Oil causes them to stick to tubes and tips, etc., and results in unacceptable losses. The following procedure is effective. Set the pipette to 50 μl, depress the plunger to the fill position, and carefully lower the tip through the oil and place it at the bottom of the tube, on or immediately above the cell pellet. Slowly release the plunger and allow the cells and the supernatant above them to enter the pipette tip in a continuous action, and continue until all or most of the supernatant has entered the tip. Keep the tip at the bottom of the tube as the level of the oil drops, and make sure no oil enters it. The aim is to flush all of the cells into the tip using the overlying supernatant as a flushing medium. The process is aided by holding the tube at eye level and carefully observing the flow of cells and supernatant in the vicinity of the pipette tip. Place the cells and supernatant in a new tube, and spin down at 16000*g* for 2.5 min. Remove the supernatant as in *Protocol 2*, step 8, and keep it.
6. Wash the cells by suspending them in 200 μl of PBS-G (*Protocol 1*) and spinning again at 16000*g* for 2.5 min in the microcentrifuge.[c] Remove all the supernatant as in *Protocol 2*, step 8, and resuspend the cells in 10 μl of PBS-G. This forms the template for the next PCR.
7. Prepare a second PCR mix as in step 1, but without the template cells (hence a total volume of 40 μl).[d] Use primers which are nested (3′) with respect to the first PCR primers, and which also include 5′ sequences designed to produce cDNA tags complementary to the fluorescent oligonucleotide(s) to be used in the third PCR. If only one target gene is being amplified it is simpler to use a single primer in the third PCR, and therefore the tagging sequence on both second PCR primers should be identical, i.e. the same nucleotide sequence as that of the fluorescent primer itself (see *Table 2* for examples).
8. Add the 10 μl of template cells from step 6 to the second PCR mix, and suspend them by pipetting.
9. Subject the tubes to 30 PCR cycles as follows:
   - denaturation: 95°C for 1 min
   - annealing: 65°C for 1 min[e]
   - extension: 72°C for 2 min
10. Wash the cells in 200 μl of PBS-G and resuspend them in a final volume of 10 μl, as in steps 4–6.

Protocol 3. *Continued*

11. Prepare a third PCR mix[d] as in step 1, but without the template cells, and using the fluorescent primer(s). If only a single fluorescent primer is being used, either add 5 μl at a concentration of 10 pmol/μl, or add 2.5 μl of primer and an extra 2.5 μl of water, to make a total volume of 40 μl.
12. Add the 10 μl of template cells from step 10 to the third PCR mix, and suspend them by pipetting.
13. Subject the tubes to 30 PCR cycles as follows:
    - denaturation: 95°C for 1 min
    - annealing: 5°C below T_m of the primers[b] for 1 min
    - extension: 72°C for 2 min
14. Wash the cells as in steps 4–6, but resuspend them in a volume of PBS-G suitable for mounting on microscope slides (e.g. 100 μl). A 1:1 mix of PBS-G:glycerol can be used as an alternative to PBS-G for fluorescence microscopy, for easier handling. Place a drop (30–50 μl) of suspension on the slide and carefully lower a coverslip over it. Seal the edges of the coverslip with nail varnish. For flow cytometry, suspend the cells in 0.5 ml of PBS-G.
15. Examine the cells for fluorescence using the appropriate excitation wavelength for the fluorochrome in the labelled primer(s).

[a] Fixed and permeabilized cells are stored in PBS-G, before and after first-strand synthesis. For use in PCR they can be washed in water before adding to the mix if preferred, but in practice the PCRs work equally well if they are added in the storage buffer. The protocol therefore specifies PBS-G throughout. If amplification problems occur, washing in water is an option which can be tried.

[b] The T_m (melting temperature) of the oligonucleotide depends on its length and its composition. A rough calculation of T_m can be made by adding together 2 for each A or T, and 4 for each C or G in the sequence. For optimum specificity, the annealing temperature should be as high as possible, ideally 5°C lower than the T_m, except where the T_m exceeds 70°C, when 65°C is high enough.

[c] More than a single wash is not recommended, but it must be stringent. Although cDNA does not appear to be released in significant quantities following first-strand synthesis (*Protocol 2*), much of the PCR product is released into the supernatant surrounding the cells, probably because it is smaller in size and because the denaturation temperature is high enough to promote release. This extracellular product must be removed. However, in permeabilized cells the effect of washing extends to the outer cytoplasm in addition to the cell surface, and excessive washing could remove the cytoplasmic PCR product which is intended as the template for the second PCR. At the level of each individual cell, washing is thus a fine balance between removing unwanted products (i.e. those originating from surrounding cells) from the cell surface, plus any which might have diffused into the outer part of the cytoplasm, while retaining the products of intracellular RT-PCR deeper in the cytoplasm.

[d] For speed and convenience, the second (and third) PCR mixes, minus the template cells, can be prepared at the same time as the first PCR mix and stored on ice or at 4°C until required. It is easily possible to run a three-PCR protocol in a working day, so *Taq* polymerase activity will not be compromised.

[e] The inclusion of tagging sequences will almost certainly result in the T_m of the second PCR primers being greater than 70°C. An annealing temperature of 65°C is sufficiently high.

Table 2. Nucleotide sequences of primers used for experiments in *Figure 1* and *Table 1*

Name of primer	Nucleotide sequence (5′ to 3′)
J_HFOR1[a]	TAGACTCACCTGCAGAGACAGTG
V_HBK1	TGCAGCTGGTGGAGTCTGGGGG
V_HBK2[a]	CAGCTCCAACTGCAGCAGCCTG
J_HLINK1[b]	CCACTGCCGCCACCACCGCTACCACCACCACCACCTGCAGAGACACTGACCAG
J_HLINK2	CCACTGCCGCCACCACCGCTACCACCACCACCAAGACTGTGAGAGTGGTGC
V_κLINK	GCGCTGCTGGCGGCAGTGGCGGCGGCGGCTCTCAAATTGTTCTCACCCAGTCTCCAGC
V_λLINK	GCGCTGCTGGCGGCAGTGGCGGCGGCGGCTCTCAGGCTGTTGTGACTCAGGAATCTGC
J_κFOR	CTTACGTTTCAGCTCCAGCTTGG
J_λFOR	GCCTAGGACAGTCAGTTTGGTTC
V_HNES1[c]	TGTAAAACGACGGCCAGTACGCTGGAGGGTCCGGAAAC
V_κNES	TGTAAAACGACGGCCAGTCCCAGCACCGAACGTGAGTGG
V_HNES2	CAGGAAACAGCTATGACCGAGCTTGTGAAGCCTGGGGCT
V_λNES	CAGGAAACAGCTATGACCCCACCGAACACCCAATGGTTGCT
ABLC	TTATCTCCACTGGCCACAAA
BCRA	AGTTACACGTTCCTGATCTC
−21BCRB[c]	TGTAAAACGACGGCCAGTTCTGACTATGAGCGTGCAGA
−21ABLD	TGTAAAACGACGGCCAGTAGTGCAACGAAAAGGTTGGG
−21M13ROX[d]	TGTAAAACGACGGCCAGT
M13 reverse ROX[d]	CAGGAAACAGCTATGACC

[a]Primers with the suffix 1 are specific for NQ10/12.5 cells, and those with suffix 2 are specific for B1-8.
[b]Linker sequences in first PCR primers are underlined. Complementary sequences result in self-assembly.
[c]Tagging sequences in the second PCR primers are underlined.
[d]The third PCR primers are red fluorescent primers obtained from Applied Biosystems (Perkin–Elmer).

Protocol 4. Electrophoresis of PCR products in agarose gels

Equipment and reagents

- Electrophoresis apparatus suitable for running submarine agarose mini-gels (gel former tray, combs, tank, and power pack)
- TAE buffer: to make a 50 × stock, dissolve 242 g Tris base and 57.1 ml of glacial acetic acid in a final volume of 900 ml of water, and add 100 ml of 0.5 M EDTA pH 8.0
- Ethidium bromide (stock solution 10 mg/ml in water)
- Agarose for gel electrophoresis
- 6 × DNA loading buffer: 30% (v/v) glycerol and 0.25% (w/v) Bromophenol Blue (or Orange G if the expected product is ≤300 bp) in TE buffer (10 mM Tris–HCl, 1 mM EDTA in water), pH 7.4, or TAE buffer
- Ultraviolet transilluminator
- DNA size marker (e.g. φX 174 DNA digested with *Hae*III), diluted to 50 ng/μl in TE buffer

Method

1. Weigh the appropriate amount of agarose[a] for each 100 ml of TAE buffer and dissolve it in the buffer by boiling; this can be done in a microwave oven, but use a bottle with a loose cap and do not fill it more than half full.
2. When completely dissolved, allow the agarose to cool to about 60°C and add 5 μl of ethidium bromide stock solution per 100 ml of agarose. Mix by swirling. Wear protective gloves (e.g. latex) when handling ethidium bromide or gels containing it.

Protocol 4. *Continued*

3. Allow any bubbles to disperse, and pour the agarose into a gel former tray appropriate for the electrophoresis tank to be used, inserting combs to cast wells allowing for a 20–25 μl sample volume.
4. Let the agarose solidify at room temperature for at least 30 min, and place the gel in the electrophoresis tank.
5. Add sufficient TAE buffer to cover the electrodes and the gel.
6. Mix 3 μl of 6 × loading buffer with 15 μl of PCR supernatant. Mix 3 μl of 6 × loading buffer with 10 μl of size marker. Add the size marker to one lane of the gel and each supernatant sample to other lanes, by carefully pipetting the sample into the well, below the surface of the TAE buffer.
7. Electrophorese the samples at about 80–100 V and 60–70 mA, until the dye front has moved 3–4 cm down the gel. This will take about 30–45 min. Examine the gel on a UV transillumінator wearing an ultra-violet face visor. Compare the band from the PCR product with bands in the size marker to determine the size of the amplified DNA fragments.

[a] For PCR products of 300–1500 bp prepare 1.5% (w/v) agarose, and for products of <300 bp prepare 2.0% agarose.

5. *In situ* PCR with biotinylated primers (see *Protocol 5*)

Protocol 5 is a two-PCR protocol, using biotin-labelled primers, which offers an alternative to fluorescent labelling. The primers used in the second PCR are again nested for optimum specificity, and incorporate 5′ biotin sites for detection by staining with streptavidin–horseradish peroxidase conjugate and a peroxidase substrate. For the post-PCR staining step, the cells are transferred to a gelatin-coated microscope slide for ease of handling. Staining could, in principle, be carried out on the cell suspension, but the repeated washings necessary to remove the staining reagents might result in an unacceptable loss of cells. Gelatin coating of the slides is essential to promote adherence of the fixed cells.

Protocol 5. *In situ* PCR with biotinylated primers for histochemical staining

Equipment and reagents

- Microcentrifuge
- Adjustable manual pipettes and sterile tips (or preferably aerosol-resistant filter tips)
- Thermal cycler (see *Protocol 3*)
- 0.5 ml Eppendorf or Sarstedt tubes
- PBS-G (see *Protocol 1*)

- Sterile distilled water
- dNTP mix, 5 mM each dNTP (see *Protocol 2*)
- Oligonucleotide primers at 10 pmol/μl
- *Taq* DNA polymerase and buffer (see *Protocol 3*)
- Fixed and permeabilized cells containing cDNA template from *Protocol 2*
- Wash bottle
- Staining jars or a slide rack and small sandwich box
- Orbital shaker
- Ethanol solutions: 50%, 70%, and 90% (v/v) in water, and 100%
- Xylene
- DPX mountant and glass coverslips
- Hydrogen peroxide, 30% (v/v) solution
- 3,3′-Diaminobenzidine tetrahydrochloride (DAB, Sigma), prepared freshly by dissolving 5 mg of DAB in 10 ml Tris–HCl pH 7.6; immediately before use add 20 μl of hydrogen peroxide. Wear protective gloves.
- Streptavidin–horseradish perxidase (streptavidin–HRP) conjugate (Dako)
- Haematoxylin stain
- Tris–HCl, pH 7.6: prepare a stock solution of 0.2 M Tris, and another of 0.1 M HCl; to 12 ml of 0.2 M Tris, add 19 ml of 0.1 M HCl and 19 ml of distilled water; check the pH and adjust if necessary
- Tris–saline, pH 7.6: dissolve 8.1 g NaCl, 0.61 g Tris, and 3.8 ml of 1 M HCl in distilled water and make up to 1 litre
- Gelatin solution for coating slides: dissolve 2.5 g of gelatin in 500 ml of distilled water, warming and stirring continuously; add 250 mg of chrome alum (chromic potassium sulfate)
- Glass microscope slides coated in gelatin by warming the gelatin solution to 50°C and incubating clean slides in it for 2–3 min; drain the slides and allow them to dry
- Humidified container, e.g. sandwich box
- Conventional light microscope

Method

1. Perform a first PCR, proceeding as in *Protocol 3*, steps 1–6. The final product will be 10 μl of cells, forming the template for the second PCR.
2. Prepare a second PCR mix as in *Protocol 3* step 1, but without the template cells (hence a total volume of 40 μl). Use primers which are nested with respect to those in the first PCR, and which are biotinylated at their 5′-end.
3. Add the 10 μl of template cells to the second PCR mix, and suspend them by pipetting.
4. Subject the cells to 30 PCR cycles as follows:
   - denaturation: 95°C for 1 min
   - annealing: 5°C below the T_m of the primer for 1 min
   - extension: 72°C for 2 min
5. After the second PCR, wash the cells in 200 μl of PBS-G as in *Protocol 3*, steps 4 and 5, and resuspend them in 200 μl of PBS-G.
6. Pipette a 50 μl aliquot on a gelatin-coated slide and spread the cells carefully into a smear without removing the coating. Allow the smear to dry for 30 min.
7. Place 200 μl of streptavidin–HRP conjugate, diluted 1/100 in Tris–saline, on the specimen. Incubate for 30 min at room temperature in a humidified container.
8. Rinse the slide gently with Tris–saline from a wash bottle. Place the slide in Tris–saline in a rack or staining jar, and agitate it gently on an orbital shaker for 3 min.

Protocol 5. *Continued*

9. Remove the slide and add 200 μl of freshly prepared DAB solution plus hydrogen peroxide to the specimen and incubate for 5 min at room temperature.
10. Rinse and wash the slide as in step 8.
11. Counter-stain by immersing the slide in haematoxylin for 30 sec, and rinse in running water for 5 min.
12. Progressively dehydrate the specimen by 3 min incubations successively in 50%, 70%, 90%, and 100% ethanol, clear it by 3 min immersion in xylene, and mount it in DPX and a coverslip.
13. Examine the cells on a conventional light microscope. Nuclei of all cells will be counter-stained blue, and amplified cDNA is seen as brown cytoplasmic staining.

References

1. Embleton, M.J., Gorochov, G., Jones, P.T., and Winter, G. (1992). *Nucleic Acids Res.*, **20**, 3831.
2. Staecker, H., Cammer, M., Rubinstein, R., and Van der Water, T.R. (1994). *Biotechniques*, **16**, 76.
3. Nuovo, G.J., MacConnell, P., Forde, A., and Delvenne, P. (1991). *Am. J. Pathol.*, **139**, 847.
4. Long, A.A., Komminoth, P., Lee, E., and Wolfe, H.J. (1993). *Histochemistry*, **99**, 151
5. Tsongalis, G.J., McPhail, A.H., Lodge-Rigal, R.D., Chapman, J.F. and Silverman, L.M. (1994). *Clin. Chem.*, **40**, 381.
6. Nuovo, G.J., Gallery, F., Hom, R., MacConnell, P., and Bloch, W. (1993). *PCR Methods Appl.*, **2**, 305.
7. Griffiths, C.M., Berek, C., Kaartinen, M., and Milstein, C. (1987). *Nature*, **312**, 272.
8. Cumano, A. and Rajewsky, K. (1985). *Eur. J. Immunol.*, **15**, 512.
9. Heisterkamp, N., Stam, K., Groffen, J., de Klein, A., and Grosveld, G. (1985). *Nature*, **315**, 758.

6

Primed *in situ* DNA synthesis (PRINS)

JOHN R. GOSDEN, ERNST J. M. SPEEL, and
YOSHIRO SHIBASAKI

1. Introduction

Techniques for visualizing DNA sequences on chromosomes have come a long way since Pardue and Gall (1,2) and John *et al.* (3) first showed that repeated sequences could be visualized in nuclei or on chromosomes by isotopic *in situ* hybridization. In those first experiments, the probe (RNA or DNA) was made radioactive with ^{3}H by growing cells in the presence of labelled precursors, purifying the nucleic acid, and hybridizing it *in situ* to cytological preparations. Later came the use of *Escherichia coli* RNA polymerase to make a complementary RNA from a known DNA template (4,5). Refinements of this process, and the introduction of cloned DNAs labelled by nick translation, permitted the localization of single-copy genes (6). However, because all these methods used autoradiography of either ^{3}H- or ^{125}I-labelled probes, mapping was a tedious process, involving the analysis of grain distribution over large numbers of metaphase spreads before genuine signals could be distinguished from background. It was only with the introduction of non-isotopic labelling methods and, in particular, the use of fluorescence, that cytological mapping of unique DNAs by *in situ* hybridization became a simple and standard laboratory technique (7), now referred to almost universally as FISH (fluorescence *in situ* hybridization).

Most recently, a new technique has been described (8,9) in which the labelled probe used in FISH is replaced by an unlabelled primer (oligonucleotide, PCR product, or cloned DNA fragment) which is annealed to the denatured DNA of cytological preparations and extended *in situ*, with the incorporation of labelled nucleotides. This technique, primed *in situ* DNA synthesis (PRINS), can replace conventional FISH in many cases where the target sequence is repeated, and has advantages of both speed (the process from annealing to analysis can be performed in half a day) and resolution (because the primer is unlabelled, and the only label not bound to the target sequence is in the form of single nucleotides, backgrounds are usually much lower than can routinely be obtained with FISH, although some samples may have backgrounds resulting from mis-priming).

In its current state of development, PRINS is still less sensitive than FISH, and no consistent reports of detection of single-copy sequences have yet been published. However, as a quick and convenient method for identifying human chromosomes by their specific alpha satellite sequences, for analysing the human chromosome complement of hybrid cell lines, as a way of analysing chromosome structure in the region of specific single-copy sequences (identified by FISH), and as an approach to investigating the evolution of repeated DNA sequences in homologous chromosomes, PRINS is proving a valuable tool, and we anticipate that the further development towards locating single-copy sequences is not too far in the future.

2. Specimen preparation

A number of preparation methods can be used for PRINS where the target material is chromosomal DNA. The simplest procedure, and the one guaranteed to give the clearest results, is to use either cultured cell lines or mitogen-stimulated peripheral blood cultures, arrested at metaphase by colcemid and treated with hypotonic medium, followed by fixation in methanol:glacial acetic acid (3:1) and spreading on acid:alcohol cleaned slides. Full details of the method are given in the literature (10,11). Alternatively, slides may be prepared directly from uncultured blood samples, by treating with hypotonic KCl (0.075 M) for 10 min, followed by fixation in methanol:glacial acetic acid and spreading on cleaned slides. Whichever method of sample preparation is practised, slides should optimally be used within two weeks. Although reactions with older slides are usually successful, results may not be as clean nor signals as strong as with fresh slides. Slides prepared from cultured or uncultured cells or blood films should be stored under vacuum in a desiccator at room temperature until used. Suspensions of fixed cells in methanol:glacial acetic acid may be stored at –20°C for several months, fresh slides being prepared from the suspensions as needed.

If slides are less than a week old, morphology is better preserved if they are passed through an ethanol series (70%, 90%, 100%) at room temperature, and air-dried, before starting the PRINS reaction.

Success may also be achieved with extended chromatin preparations (see *Protocol 5*), frozen sections (see *Protocol 6*), and alternative fixation procedures for combining PRINS with immunocytochemistry (see *Protocol 7*).

3. PRINS

For any PRINS reaction, the basic reaction system is the same and is given in *Protocol 1*. Unless a flat-bed programmable thermal cycler is available, carry out chemical denaturation (see section 3.1) before proceeding with PRINS.

3.1 Denaturation

For chemical denaturation, proceed as follows:

(a) Incubate the slides in 70% formamide, 2 × SSC at 70 °C for 2 min.
(b) Transfer rapidly to ice-cold 70% ethanol for 2 min, followed by 90% and 100% ethanol.
(c) Air-dry before proceeding to the next step.

If a thermal cycler is available, the denaturation step can be integrated into the PRINS reaction (see below). Thermal cyclers with a flat bed for microscope slides are not yet widely available. Some of the products sold for this purpose are not altogether suitable, as they are *ad hoc* modifications of machines designed for PCR in microcentrifuge tubes, with a plate added to the heated block. Thermal transfer and temperature control in such a system are rarely satisfactory. The procedure can be carried out by transferring slides through a series of water baths at appropriate temperatures, but this too means that temperature control cannot be precise, and the temperature drop during the transfer from water bath to water bath leads to high backgrounds. We have found the most suitable purpose-built products to be the OmniGene In Situ and OmniSlide (Hybaid Ltd), which hold four and 20 slides respectively (see Chapter 9).

3.2 The basic PRINS procedure

This is described in *Protocol 1*.

Protocol 1. Plain PRINS

Equipment and reagents

- PRINS buffer (10×): 500 mM KCl, 100 mM Tris–HCl, pH 8.3, 15 mM $MgCl_2$, 0.1% BSA
- dATP, dCTP, dGTP, and dTTP; separate 100 mM solution of each (Pharmacia Biotech), diluted 1:10 with sterile distilled water
- 1 mM biotin-16-dUTP (Bio-16-dUTP; Boehringer Mannheim), or 1 mM digoxigenin-11-dUTP (Dig-11-dUTP; Boehringer Mannheim), or 1 mM FluoroRed, FluoroGreen, or FluoroBlue (Amersham International, plc)
- Oligonucleotide primer(s) at 250 ng/μl[a]
- *Taq* DNA polymerase: *Taq* (Boehringer), Ampli*Taq* (Cetus), or Thermoprime^{plus} (Advanced Biotechnologies Ltd)
- Rubber cement (vulcanizing solution) (e.g. Tip-Top, Stahlgruber)[b]
- Stop buffer: 500 mM NaCl, 50 mM EDTA
- Wash buffer: 4 × SSC (diluted from stock 20 × SSC), 0.05% Triton X-100
- Water bath at 65 °C

Method

1. Make up the reaction mix as follows: for each slide, put 1 μL of each of the diluted nucleotide triphosphates, plus 1 μL of the selected labelled dUTP (biotin-, digoxigenin-, or fluorochrome-conjugated), 5 μL of 10 × PRINS buffer, and 1 μL of the appropriate oligonucleotide

Protocol 1. *Continued*

primer(s)[c] into a microcentrifuge tube on ice, and add distilled water to 50 μL.

2. Mix thoroughly and add 1 unit of DNA polymerase; mix carefully and place 40 μL on a clean coverslip.
3. Pick the coverslip up with a slide (this spreads the reaction mix evenly, with least risk of introducing air bubbles) and seal with rubber cement.
4. Dry the seal (a cold-air fan is quick and safe) and transfer the slides to the flat block of a thermal cycler. A suitable basic program for the Hybaid OmniGene In Situ or Hybaid OmniSlide is:
   (a) 93°C, 3 min (denaturation)
   (b) 60°C, 5–10 min (annealing)
   (c) 72°C, 10–30 min (extension)

   If no programmable cycler is available, after formamide denaturation and dehydration (see section 3.1), place the reaction mix on the coverslip, pick it up using the slide, and seal. When the seal is dry, transfer the slides to a water bath at the appropriate annealing temperature for 10 min, then transfer to a water bath at 72°C for 15–30 min for the extension reaction.
5. On completion of the program, remove the seal (it can easily be peeled off by rubbing one corner) and transfer the slides for 1 min to a Coplin jar containing stop buffer at 65°C. Leave the coverslips in place, unless they come off readily with the seal (they will in any case fall off in the stop buffer). After 1 min, transfer the slides to a staining dish containing wash buffer. They may be kept in this solution overnight if convenient.[d]

[a] Oligonucleotide primers can be synthesized on an ABI DNA synthesizer, and used without further purification other than ethanol precipitation and washing. If this facility is not available, they may be obtained from commercial sources, but purification steps such as HPLC are not needed, and only increase the cost of the product. As an alternative, complete systems for chromosome identification by PRINS are becoming available (e.g. Advanced Biotechnologies Ltd).

[b] A suitable seal should be reasonably robust, provide a vapour-tight seal, and be easily and completely removed at the end of the procedure. 'Tip-Top', which is readily available from bicycle repair shops, fulfils all these requirements.

[c] The majority of chromosome-specific alphoid sequences produce adequate signal with a single primer at a concentration of 250 ng/50 μl reaction. In some cases a clearer signal with less background may be produced with paired primers, at the same concentration, while in others the concentration of primer may be reduced, with a concomitant reduction in cross-reaction to related chromosomal sequences.

[d] Slides which have been labelled directly with fluorochromes may still be kept in wash buffer overnight if convenient, but should be kept in the dark to prevent bleaching and fading of the label.

3.3 Detection methods

Protocol 2. Detection of fluorescence

Equipment and reagents

- Dried skimmed milk powder
- Avidin-DCS–fluorescein isothiocyanate (avidin–FITC; Vector Labs)
- Avidin-DCS–Texas Red (avidin–TR; Vector Labs)
- Anti-digoxigenin–fluorescein (anti-Dig–FITC; Boehringer Mannheim)
- Anti-digoxigenin–rhodamine (anti-Dig–rhodamine; Boehringer Mannheim)
- Propidium iodide (20 μg/ml)
- VectaShield (Vector Labs)
- 4′,6-diamidino-2-phenylindole·2HCl (DAPI) (100 μg/ml)
- 20 × SSC (3.0 M NaCl, 0.3 M trisodium citrate, pH 7.3)
- Wash buffer; 4 × SSC (diluted from stock 20 × SSC), 0.05% Triton X-100
- Incubator or water bath at 37 °C and water bath at 45 °C
- Microscope equipped for epifluorescence (e.g. Zeiss Axioskop or Leitz Ortholux II with Ploemopak filter system)

Method[a]

Note: It is important that the slides do not become dry at any time during this process.

1. Prepare blocking buffer by dissolving milk powder (5%, w/v) in wash buffer (the milk powder dissolves more rapidly if the solution is warmed to 45 °C for a few seconds).
2. Put 40 μl of blocking buffer on a clean coverslip, shake surplus wash buffer from slide, and pick up coverslip carrying blocking buffer. Leave (unsealed) at room temperature for 5 min.
3. Dissolve reporter (avidin–fluorochrome or anti-digoxigenin–fluorochrome) in blocking buffer. For avidin–FITC or avidin–TR, 1:500 is a suitable dilution; anti-Dig–FITC- or anti-Dig–rhodamine are better at 1:100 dilution. Make sufficient for 40 μl/slide. Spin in a micro-centrifuge for 5 min. This precipitates any aggregates which may have formed during storage and can cause high and non-specific back-ground.
4. Remove the coverslip from the slide, shake surplus fluid off both slide and coverslip, and add 40 μl of reporter solution to the same coverslip. Pick up the coverslip with the slide and incubate in a moist chamber (e.g. a sandwich box lined with damp filter paper) at 37 °C for 30–60 min.
5. Warm a reagent bottle containing wash buffer to 45 °C in a water bath. Remove coverslips, and wash the slides 3 × 2 min in wash buffer at 45 °C.
6. After the final wash, shake off surplus fluid and mount the slides in VectaShield containing the appropriate counter-stain:
   - for slides labelled with rhodamine or Texas Red, this should be DAPI (0.5 μg/100 μl VectaShield, i.e. 5 μl of DAPI stock/100 μl VectaShield)

Protocol 2. *Continued*

- for slides labelled with FITC this should be a propidium iodide/DAPI mixture (3.75 μl of each stock/100 μl VectaShield)

Use 20–30 μl mountant/slide, blot surplus by covering slide and coverslip with a tissue and pressing gently to expel excess mountant, and seal with rubber cement.[b]

[a]Steps 1–5 apply only to slides in which the PRINS reaction has been labelled with biotin or digoxigenin. Slides in which the reaction used a fluorochrome-dUTP as the label require no detection step: simply mount them in VectaShield with an appropriate counter-stain (see step 6)
[b]Slides may be stored in the dark at 4°C for several months. If the stain shows signs of fading, simply peel off the sealant, soak the slide overnight in 4 × SSC, 0.05% Triton X-100 (the coverslip will fall off at this point), and remount as above.

Protocol 3. Enzyme cytochemical detection

Equipment and reagents

- Normal goat serum (NGS)
- Horseradish peroxidase (HRP)-conjugated avidin (avidin–HRP; Dako)
- HRP-conjugated sheep anti-digoxigenin Fab fragments (sheep anti-Dig–HRP; Boehringer Mannheim)
- HRP-conjugated goat anti-mouse IgG (goat anti-mouse–HRP; Dako)
- Diaminobenzidine (DAB; Sigma)
- 3,3′,5,5′-Tetramethylbenzidine (TMB; Sigma)
- 30% Hydrogen peroxide
- Monoclonal mouse anti-digoxin antibody (mouse anti-Dig; Sigma)
- Alkaline phosphatase (AP)-conjugated avidin (avidin–AP; Dako)
- AP-conjugated swine anti-rabbit IgG antibody (swine anti-rabbit–AP; Dako)
- AP-conjugated goat anti-mouse antibody (goat anti-mouse–AP; Dako)
- Fast Red TR (Sigma)
- Rabbit anti-fluorescein antibody (rabbit anti-FITC; Dako)
- Naphthol AS-MX phosphate (Sigma)
- Dioctyl sodium sulfosuccinate (DSSS) (Sigma)
- Sodium tungstate (Sigma)
- PBS containing 0.05% Triton X-100 and 2% NGS
- HRP–DAB buffer: 0.1 M imidazole (Merck) in PBS, pH 7.6
- HRP–TMB buffer: 100 mM citrate-phosphate buffer, pH 5.1
- Polyvinyl alcohol (PVA), mol. wt 40 000 (Sigma)
- AP–Fast Red buffer: 0.2 M Tris–HCl pH 8.5, 10 mM $MgCl_2$, 5% PVA
- Haematoxylin: haematoxylin (Solution Gill No. 3; Sigma) diluted 1:4 with distilled water
- 400 ASA and 100 ASA film
- Blue and magenta filters

Method

1. Follow steps 1 and 2 from *Protocol 2*.
2 Dilute reporter as follows:
   - for reactions using biotin-16-dUTP, dilute avidin–HRP 1:100 in blocking buffer and apply 40 μl under a coverslip.[a,c,d]
   - for reactions using digoxigenin-11-dUTP, dilute mouse anti-Dig 1:2000 in PBS containing 0.05% Triton X-100 and 2% NGS.[b,d]
   - for reactions using fluorescein-12-dUTP, dilute rabbit anti-FITC 1:2000 in PBS, 0.05% Triton X-100, 2% NGS.

3. Incubate for 30 min at 37 °C in a humid chamber.
4. Wash the slides for 2 × 5 min in washing buffer (for biotin–avidin reactions) or PBS; 0.05% Triton X-100; 2% NGS (for antibody reactions).
5. Mouse anti-Dig and rabbit anti-FITC can be detected with an additional layer of goat anti-mouse–HRP or swine anti-rabbit–HRP respectively, both diluted 1:100 in PBS, 0.05% Triton X-100, 2% NGS. Incubate for 30 min at 37 °C in a humid chamber.[c,d]
6. Wash the slides for 2 × 5 min in PBS, 0.05% Triton X-100, 2% NGS
7. Visualize the PRINS-labelled DNA target using an appropriate HRP (a, b) or AP reaction (c):
   (a) HRP–DAB reaction: mix 1 ml of 5 mg/ml DAB in PBS, 9 ml HRP–DAB buffer, and 10 μl 30% hydrogen peroxide just before use and overlay each sample with 100 μl under a coverslip. Incubate the the slides for 5–15 min at 37 °C, wash for 3 × 5 min with PBS, and (optionally) dehydrate.
   (b) HRP–TMB reaction: dissolve 100 mg sodium tungstate in 7.5 ml HRP–TMB buffer, and adjust the pH of this solution to pH 5.0–5.5 with 37% HCl. Just before use, dissolve 20 mg DSSS and 6 mg TMB in 2.5 ml 100% ethanol at 80 °C. Mix both solutions with 10 μl hydrogen peroxide, and overlay each sample with 100 μl under a coverslip. Incubate the the slides for 1–2 min at 37 °C, wash for 3 × 1 min with ice-cold 0.1 M phosphate buffer (pH 6.0), and dehydrate.
   (c) AP–Fast Red reaction: mix 4 mL of AP buffer, 250 μl of AP buffer without PVA containing 1 mg naphthol AS-MX phosphate, and 250 μL of AP buffer without PVA containing 5 mg Fast Red TR just before use, and overlay each sample with 100 μL of this mix under a coverslip. Incubate the the slides for 5–15 min at 37 °C and wash for 3 × 5 min with PBS.
8. Counter-stain the samples with haematoxylin for 4–5 min, wash for 1 min in tap water and 1 min in distilled water, and, if you wish, air-dry.
9. Mount samples:
   - HRP–DAB precipitate: mount in an organic or aqueous mountant
   - HRP–TMB precipitate: mount in an organic mountant or immersion oil
   - AP–Fast Red precipitate: mount in an aqueous mountant
10. Examine the slides under a bright-field microscope. Microphotographs can be made using blue and magenta filters and 100 ASA film.

[a] Amplification of biotin-16-dUTP may be achieved by incubation with biotinylated goat anti-avidin (Vector), diluted 1:100 in blocking buffer, and again avidin–HRP.
[b] In many cases a single detection step with HRP-conjugated sheep anti-digoxigenin (Boehringer) can be used, diluting 1:100 in blocking buffer.
[c] HRP conjugates can be substituted with AP conjugates for AP–Fast Red visualization.
[d] PRINS signals may also be amplified by combining these detection systems with peroxidase-mediated deposition of hapten- or fluorochrome-labelled tyramides (12)

3.4. Variations on the basic method

3.4.1 Multiple PRINS

Each PRINS reaction can only identify one pair of homologous chromosomes, because the nature of the reaction means that the product of only one primer or primer pair can be specifically labelled in each reaction. However, by inserting a blocking step after each PRINS reaction, to ensure that the 3′-ends of the products of the previous reaction cannot act as primers for the next reaction, it is possible to perform several PRINS reactions on a single slide, and therefore ascertain the number of each of several pairs of chromosomes present in a given sample (13,14) (see *Protocol 4*).

Protocol 4. Multiple PRINS

Additional reagents

- 0.025 mM each of ddATP, ddCTP, ddGTP, and ddTTP (Pharmacia)
- DNA polymerase 1, large fragment (Klenow enzyme; Boehringer Mannheim)
- 10× nick translation buffer (10 × NT): 0.5 M Tris–HCl, pH 7.2, 0.1 M $MgSO_4$, 0.1 mM dithiothreitol, 1.5 mg/ml BSA

Method

1. Perform the first PRINS reaction as described in *Protocol 1*, using biotin or digoxigenin as the label.
2. Make up the block reaction with (for each slide) 4 μl of each of the four ddNTPs, plus 4 μl 10 × NT, and distilled water to a total of 40 μl.
3. After the stop buffer (step 5 in *Protocol 1*), rinse the slides briefly in 1 × NT buffer, and shake off the surplus fluid.
4. Add 1 unit of Klenow enzyme per slide to the block reaction, and aliquot 40 μl on to each coverslip.
5. Pick up the coverslips with the slides and incubate in a humid chamber (e.g. a plastic box containing damp filter paper) for 1 h at 37 °C.
6. Pass the slides through an ethanol series (70%, 90%, and 100%) and air-dry before starting the second PRINS reaction.
7. For subsequent PRINS reactions:
   (a) Make up a standard PRINS reaction mix with a different primer and reporter (e.g. use digoxigenin for the first PRINS, biotin for the second, and FluoroRed for the third) and perform annealing and extension stages of the reaction, omitting the denaturation step.[a]
   (b) Stop the reaction as before and, if a third PRINS is required, go through the blocking process again (steps 2–6 above) before carrying out the third PRINS.
   (c) Otherwise, transfer the slides to 4 × SSC, 0.05% Triton X-100.

8. Detect fluorescence as described in *Protocol 2*, adding fluorochrome-conjugated anti-digoxigenin and avidin to the slides simultaneously. Alternatively, perform enzyme-based detection, combining the HRP–DAB, AP–Fast Red, and HRP–TMB reactions (see *Protocol 3* and references 15 and 16).

[a] No denaturation step is needed after the first, as the chromosomal DNA appears to remain single-stranded throughout the PRINS incubations.

3.4.2 PRINS on extended chromatin

A problem with conventional FISH on metaphase chromosomes has been that it is difficult to resolve closely associated markers. Any targets separated by less than about 1 Mb (1×10^6 bp, about the size of a chromosome band) tend to appear as a single locus on metaphase chromosomes. Attempts to solve this problem have included hybridization to interphase nuclei (17,18), in which, because the chromatin is decondensed, the markers are further apart, and therefore more easily resolved.

Recently a number of methods have been described that permit the generation of free, partly decondensed chromatin fibres (19–23). FISH can be performed on this material with much greater resolution than on metaphase chromosomes and, because the path of the chromatin fibres is generally linear and can be readily traced, mapping physical distances and establishing linkage is relatively easy, and markers can be mapped when they are separated by as little as 3–5 kb. PRINS provides an alternative approach to this material. This technique can be usefully applied to extended chromatin both for structural analysis of chromatin organization, and, by combining FISH with PRINS (see *Protocol 5*), for relating mapped markers to specific chromatin organization patterns (16).

Protocol 5. PRINS on extended chromatin

Reagents

- Lysis solution: 50 mM NaOH, 30% ethanol
- Phosphate-buffered saline (PBS)
- Methanol
- Ethanol series: 70%, 95%, and 100%

Method

1. In the centre of a cleaned glass slide, drop 20 μL of fixed cell suspension.[a]
2. Before the drop has completely dried, immerse the slides in PBS in a Coplin jar and incubate for 1 min. This procedure neutralizes the acid fix in which the cells are suspended.
3. Blot excess by shaking the slide and touching the edge to filter paper, and pipette 100 μL of lysis solution on the slide held horizontally followed by 100 μL of methanol gently on top of the lysis solution. The

Protocol 5. *Continued*

rapid change of surface tension draws out the chromatin fibres from the lysed cells.

4. Blot excess fluid as above, and air-dry the slides. When they are dry, pass the slides through an ethanol series from 70% to 100% and air-dry.
5. Place the slides in an incubator at 80 °C for 1 h and allow to cool. These slides should be used within a few days of preparation.
6. Thereafter, perform the PRINS reaction exactly as described in section 3.1 and *Protocol 1*.

[a] Stored cultured metaphase chromosome preparations kept in methanol:acetic acid at –20 °C (see section 2) can be used for up to 1 year.

3.4.3 PRINS on frozen sections

The PRINS labelling reaction can also be applied to frozen tissue sections (24) (see *Protocol 6*). In this way individual cells can be identified in their tissue context and be analysed for their copy numbers of specific chromosome regions. Both tissue fixation and proteolytic digestion before performing the PRINS reaction are critical steps in the total procedure, permitting access of the PRINS reactants while preserving the morphology of the nuclei in the tissue. Such pre-treatment steps have also been shown to be essential for efficient application of *in situ* hybridization to tissue sections (25,26). The technique can be successfully combined with fluorescence as well as bright-field analysis. In the latter case absorption stain end-points are used for the detection procedure, as an alternative to fluorescent markers (see *Protocols 2* and *3*).

Protocol 6. PRINS on frozen sections

Additional equipment and reagents

- Liquid nitrogen
- Cryostat
- Poly-L-lysine-coated slides
- Methanol:acetic acid (3:1)
- PBS
- 0.01 M HCl
- Pepsin from porcine stomach mucosa (2500–3500 U/mg) (Sigma), 100 μg/ml in 0.01 M HCl
- 1% Paraformaldehyde in PBS

Method

1. Using liquid nitrogen, snap-freeze fresh tissue samples obtained after surgical resection.
2. From each sample, cut 4 μm sections with a cryostat, mount them on poly-L-lysine-coated slides, and store at –20 °C until use.
3. Air-dry the slides, fix in methanol:acetic acid (3:1) for 10 min at room temperature, and air-dry again.[a]

4. Wash the slides for 5 min in PBS and for 2 min in 0.01 M HCl.
5. Treat samples with 100 μg/ml pepsin in 0.01 M HCl for 10 min at 37 °C, wash for 2 min in 0.01 M HCl at 37 °C, and pass the slides through an ethanol series starting with 70% ethanol in 0.01 M HCl.[b]
6. Post-fix samples in 1% paraformaldehyde in PBS for 20 min at 4°C, wash in PBS for 5 min, dehydrate, and subject the slides to the PRINS procedure.
7. Prepare the PRINS reaction mix on ice as in section 3.1 and *Protocol 1*, and perform the reaction as described there.

[a]Fixation of frozen tissue sections with other fixatives, such as acetone (10 min, –20°C), methanol (10 min, –20 °C), methanol then acetone (1 min, –20 °C/3 × 5 sec, room temperature), 70% ethanol (10 min, –20 °C), or 70% ethanol, 1% formaldehyde (10 min, –20 °C), results in poor preservation of cell morphology after PRINS. In addition, we frequently observe fluorescent staining of the entire nucleus after PRINS labelling of tissue sections fixed in methanol–acetone, probably caused by nuclease activities that survive methanol–acetone fixation.
[b]Dehydration of the samples after pepsin treatment starting with 70% ethanol in 0.01 M HCl helps to preserve cell morphology.

3.4.4 Combined PRINS and immunocytochemistry

The combination of PRINS and immunocytochemistry makes it possible, for example, to immunophenotype cells with a specific chromosomal content or viral infection. The success and sensitivity of such a combined procedure depends on factors such as preservation of cell morphology and protein epitopes, accessibility of nucleic acid targets, lack of cross-reaction between the different detection procedures, good colour separation, and stability of fluorochromes and enzyme cytochemical precipitates. Since several steps in the PRINS procedure (enzymatic digestion, post-fixation, denaturation at high temperatures) may destroy antigenic determinants, a procedure starting with immunocytochemistry and followed by PRINS is usually preferred. A variety of procedures have already been reported for the combination of immunocytochemistry and ISH (for a review, see 27).

Protocol 7 describes the application of a sensitive, high resolution fluorescence alkaline phosphatase (AP)–Fast Red immunocytochemical staining method (28) in combination with subsequent PRINS labelling of DNA target sequences in cell preparations. The slow-fading property of the AP–Fast Red precipitate as well as its stability during enzymatic pre-treatment steps and the entire PRINS procedure are essential for accurate immunostaining. As a model system, somatic cell hybrid and tumour cell lines have been used for simultaneous detection of surface antigens (epidermal growth factor receptor (EGFR) and neural cell adhesion molecule (N-CAM)) and repeated chromosome-specific DNA sequences.

Protocol 7. PRINS combined with immunocytochemistry

Additional reagents

- Normal goat serum (NGS)
- Monoclonal antibody EGFR1, directed against EGFR (a gift from Professor V. van Heyningen, Edinburgh, UK)
- Monoclonal antibody 163A5, directed against a cell-surface marker of J1Cl4 cells (29)
- Monoclonal antibody RNL1, directed against N-CAM (30)
- Naphthol AS-MX phosphate (Sigma)
- Fast Red TR (Sigma)
- AP-conjugated goat anti-mouse IgG (goat anti-mouse–AP) (Dako)
- Polyvinyl alcohol (PVA), mol. wt 40 000 (Sigma)
- AP buffer: 0.2 M Tris–HCl, pH 8.5, 10 mM $MgCl_2$, 5% PVA
- Blocking buffer: PBS (diluted from stock 10 × PBS), 0.05% Triton X-100, 2–5% NGS
- Washing buffer: PBS, 0.05% Triton X-100

Method

1. Culture hybrid (Cl21-TN6, J1C14) or tumour (H460) cell lines on glass slides by standard methods (29,31,32).
2. Fix in cold methanol (–20 °C) for 5 sec then cold acetone (4 °C) for 3 × 5 sec, air-dry, and store at –20 °C until use.[a]
3. Incubate the slides for 10 min at room temperature with blocking buffer.
4. Incubate the slides for 45 min at room temperature with undiluted culture supernatant of the appropriate antigen-specific monoclonal antibody containing 2% NGS.
5. Wash the slides for 2 × 5 min with washing buffer.
6. Incubate the slides for 45 min at room temperature with goat anti-mouse–AP, diluted 1:50 in blocking buffer.[b]
7. Wash the slides for 5 min with washing buffer, and for 5 min with PBS.
8. Visualize the antigen with the AP–Fast Red reaction:
   (a) Mix together 4 ml AP buffer, 1 mg naphthol AS-MX-phosphate in 250 μL buffer without PVA, and 5 mg Fast Red TR in 750 μl buffer without PVA just before use.
   (b) Overlay each sample with 100 μl under a coverslip.
   (c) Incubate the slides for 5–15 min at 37 °C and wash for 3 × 5 min with PBS.[c]
9. Process cells for PRINS as follows:
   (a) Wash the slides for 2 min at 37 °C with 0.01 M HCl.
   (b) Incubate the samples with 100 μg/ml pepsin in 0.01 M HCl for 20 min at 37 °C.
   (c) Wash again with 0.01 M HCl for 2 min.

(d) Post-fix the slides in 1% paraformaldehyde in PBS for 20 min at 4°C.

(e) Wash cells in PBS for 5 min at room temperature, followed by a wash in PRINS buffer for 5 min at room temperature.

10. Thereafter proceed as described in *Protocol 1*, and either *Protocol 2* or *3*. If the PRINS label is detected with a fluorescence-based method (e.g. FITC), counter-stain with DAPI and mount in VectaShield (see *Protocol 2*).

[a] Since methanol–acetone fixation is a very mild procedure, preservation of cell morphology may be a serious problem after PRINS. Furthermore, we frequently observe fluorescent staining of the entire nucleus after PRINS labelling, probably due to nicking of chromosomal DNA caused by nuclease activities that survive methanol–acetone fixation. Therefore, other fixatives should be tested that are compatible with antigen detection but preserve cell morphology better while still permitting specific PRINS labelling. In the case of H460 cells, fixation with cold 70% ethanol (–20°C) for 10 min is a valid alternative to methanol/acetone.
[b] If amplification of the immunocytochemical signal is needed, a third detection step may be added after this second incubation step.
[c] Monitor the enzymic reaction under the microscope to adjust the reaction time so that the precipitate becomes discretely localized and not so dense that it masks nucleic acid sequences in the PRINS reaction. To ensure the specificity of the AP–Fast Red staining, a control slide with FITC-conjugated secondary antibodies is recommended for comparison. Staining specificity can be lost if cells contain endogenous AP activity. This endogenous enzyme can be inhibited by the addition of levamisole (Sigma) to the reaction medium to a final concentration of 1–5 mM. Do not dehydrate the slides after the AP reaction, as the precipitate dissolves in organic solvents. Optionally, you may air-dry the slides after rinsing in distilled water.

References

1. Gall, J. and Pardue, M.L. (1969). *Proc. Natl Acad. Sci. USA*, **63**, 378.
2. Pardue, M.L. and Gall, J. (1969). *Proc. Natl Acad. Sci. USA*, **64**, 600.
3. John, H.A., Birnstiel, M.L., and Jones, K.W. (1969). *Nature*, **223**, 582.
4. Pardue, M.L. and Gall, J.G. (1970). *Science*, **168**, 1356.
5. Jones, K.W. and Corneo, G (1971). *Nature New Biology*, **233**, 268.
6. Harper, M.E., Ullrich, A., and Saunders, G.F. (1981). *Proc. Natl Acad. Sci. USA*, **78**, 4458.
7. Pinkel, D., Straume, T., and Gray, J.W. (1986). *Proc. Natl Acad. Sci. USA*, **83**, 2934.
8. Koch, J.E., Kølvraa, S., Petersen, K.B., Gregersen, N., and Bolund, I. (1989). *Chromosoma*, **98**, 259.
9. Gosden, J., Hanratty, D., Starling, J., Fantes, J., Mitchell, A., and Porteous, D. (1991). *Cytogenet. Cell Genet.*, **57**, 100.
10. Spowart, G. (1994). In *Methods in molecular biology*, vol. 29: *Chromosome analysis protocols* (ed. J.R. Gosden), p. 1. Humana Press, Totowa, NJ.
11. Fletcher, J. (1994). In *Methods in molecular biology*, vol. 29: *Chromosome analysis protocols* (ed. J.R. Gosden), p. 51. Humana Press, Totowa, NJ.
12. Bobrow, M.N., Harris, T.D., Shaughnessy, K.J., and Litt, G.J. (1989). *J. Immunol. Methods*, **125**, 279.

13. Gosden, J. and Lawson, D. (1994). *Hum. Mol. Genet.*, **3**, 931.
14. Speel, E.J.M., Lawson, D., Hopman, A.H.N., and Gosden, J. (1995). *Hum. Genet.*, **95**, 29.
15. Speel, E.J.M., Jansen, M.P.H.M., Ramaekers, F.C.S., and Hopman, A.H.N. (1994). *J. Histochem. Cytochem.*, **42**, 1299.
16. Speel, E.J.M., Lawson, D., Ramaekers, F.C.S, Gosden, J.R., and Hopman, A.H.N. (1996). In *Methods in molecular biology: protocols for PRINS and in situ PCR* (ed. J.R. Gosden). Humana Press, Totowa, NJ., p. 53.
17. Trask, B.J. (1991). *Trends Genet.*, **7**, 149.
18. Van den Engh, G., Sachs, R., and Trask, B. (1992). *Science*, **257**, 1410.
19. Heng, H.H.Q., Squire, J., and Tsui, L.-C. (1992). *Proc. Natl Acad. Sci. USA*, **89**, 9509.
20. Wiegant, J., Kalle, W., Mullenders, L., Brookes, S., Hoovers, J.M.N., Dauwerse, J.G., van Ommen, G.J.B., and Raap, A.K. (1992). *Hum. Mol. Genet.*, **1**, 587.
21. Parra, I. and Windle, B. (1993). *Nature Genet.*, **5**, 17.
22. Heiskanen, M., Karhu, R., Hellsten, E., Peltonen, L., Kallionemi, O.P., and Palotie, A. (1994). *BioTechniques*, **17**, 928.
23. Senger, G., Jones, T.A., Fidlerová, H., Sanséau, P., Trowsdale, J., Dutt, M., and Sheer, D. (1994). *Hum. Mol. Genet.*, **3**, 1275.
24. Speel, E.J.M., Lawson, D., Ramaekers, F.C.S., Gosden, J.R., and Hopman, A.H.N. (1996). *BioTechniques*, **20**, 226.
25. Hopman, A.H.N., Van Hooren, E., Van der Kaa, C.A., Vooijs, G.P., and Ramaekers, F.C.S. (1991). *Modern Pathol.*, **4**, 503–513.
26. Hopman, A.H.N., Poddighe, P.J., Moesker, O., and Ramaekers, F.C.S. (1992). In *Diagnostic molecular pathology: a practical approach*, vol. 1 (ed. C.S. Herrington and J.O'D. McGee), pp. 141–167. IRL Press, Oxford.
27. Speel, E.J.M., Ramaekers, F.C.S., and Hopman, A.H.N. (1995). *Histochem. J.*, **27**, 833.
28. Speel, E.J.M., Herbergs, J., Remaekers, F.C.S., and Hopman, A.H.N. (1994). *J. Histochem. Cytochem.*, **42**, 961.
29. Glaser, T., Housman, D., Lewis, W.H., Gerhard, D., and Jones, C. (1989). *Somat. Cell Mol. Genet.*, **15**, 477.
30. Boerman, O.C., Mijnheere, E.P., Broers, J.L.V., Vooijs, G.P., and Ramaekers, F.C.S. (1991). *Int. J. Cancer*, **48**, 457.
31. Dorin, J.R., Inglis, J.D., and Porteous, D.J. (1989). *Science*, **243**, 1357.
32. Carney, D.N., Gazdar, A.F., Bepler, G., Guccion, J.G., Marangos, P.J., Moody, T.W., Zweig, M.H., and Minna, J.D. (1985). *Cancer Res.*, **45**, 2913.

7

Application of *in situ* PCR techniques to human tissues

OMAR BAGASRA, LISA E. BOBROSKI, MUHAMMED AMJAD, ROGER J. POMERANTZ, and JOHN HANSEN

1. Introduction

Since we first reported our findings regarding *in situ* amplification of the HIV-1 *gag* gene in an HIV-1-infected cell line in March 1990 (1), there has been an explosion of research in the area of *in situ* polymerase chain reaction (*in situ* PCR). There are over 300 publications describing various forms of *in situ* gene amplifications (reviewed in 2,3,13) , identifying various infectious agents, tumour marker genes, cytokines, growth factors and their receptors, and other genetic elements of interest, in peer-reviewed journals (1–25). The solution-based PCR method for amplifying defined gene sequences has proved a valuable tool not only for basic researchers but also for clinical scientists. Using even a minute amount of DNA or RNA and choosing a thermostable enzyme from a large variety of sources, the gene of interest can be amplified and then analysed and/or sequenced. Thus genes or segments of gene sequences present only in a small sample of cells or a small fraction of mixed cellular populations can be examined. However, one of the major drawbacks of solution-based PCR is that it does not allow the association of amplified signals of a specific gene segment with the histological cell type(s) (13,14). For example, it would be advantageous to determine what types of cells in the peripheral blood carry aberrant genes in a leukaemia patient and what percentage of leukaemia cells is present after various forms of anti-tumour therapy.

The ability to identify individual cells expressing or carrying specific genes of interest in a tissue section, under the microscope, provides a great advantage in determining various aspects of normal, as opposed to pathological, conditions. For example, this technique could be used in determination of tumour burden, before and after chemotherapy, in lymphomas or leukaemias, where specific aberrant gene translocations are associated with certain types of malignancy (23). In the case of HIV-1 or other viral infections, the effects of therapy or putative anti-viral vaccination can be determined by evaluating the number of cells still infected with viral agent after chemotherapy or vaccination (2,5,6,20). Similarly, potentially pre-neoplastic lesions can be identified by examining tumour suppressor genes, for example p53 mutations associated with certain tumours, or oncogenes or other aberrant gene sequences which are known to be associated with certain types of tumours (23). In the area of diagnostic pathology, determination of origin of metastatic tumours is a perplexing problem. By using the proper primers for genes which are expressed by certain histological cell types, the origin of metastatic tumours can potentially be determined by performing reverse transcriptase (RT)-initiated *in situ* PCR (23).

Our laboratories have been using *in situ* PCR techniques for several years and we have developed a simple, sensitive *in situ* PCR which has proved reproducible in multiple double-blinded studies (13). This method can be used for amplifying both DNA and RNA gene sequences. By using multiply labelled probes, various signals can be detected in a single cell. In addition, under special circumstances, immunohistochemistry, RNA and DNA amplification can be performed at a single cell level (the so-called 'triple labelling') (14).

To date, we have successfully amplified and detected HIV-1, simian immunodeficiency virus, human papillomavirus, hepatitis B virus, cytomegalovirus, Epstein–Barr virus, human herpes virus 6 and 8, herpes simplex virus, lymphogranuloma venereum, p53 and its mutations, mRNA for surfactant Protein A , oestrogen receptors, inducible nitrous oxide synthesis gene sequences associated with multiple sclerosis, by DNA and/or RNA (RT *in situ* PCR), in various tissues, including: peripheral blood mononuclear cells (PBMCs), lymph nodes, spleen, skin, breast, lungs, placenta, sperm, cytological specimens, tumours, cultured cells, numerous other formalin-fixed, paraffin-embedded tissues, and various cellular populations in the frozen sections of AIDS brains, infected with HIV-1 (1–14,17,18,22–25). In the following pages, a moderately detailed protocol, currently being used in our laboratory, will be presented. For more details consult the references cited at the end of the chapter.

2. Preparation of glass slides and tissues

Before *in situ* reactions can be performed by this protocol, the proper sort of glass slide with Teflon coating must be obtained and the glass surface treated with the appropriate silicon compound. Both of these factors are very important, as detailed in the following two sections.

2.1 Glass slides with Teflon-edged wells

Glass slides should always be used. Not only is glass able to withstand the stress of repeated heat denaturation, but also its chemical surface, namely silicon oxide, is suitable for proper silanization. Slides with special Teflon coatings that form individual 'wells' are useful because vapour-tight reaction chambers can be formed on the surface of the slides when coverslips are applied with coatings of nail varnish around the edge. These reaction chambers are necessary because within them, proper tonicity and ion concentrations can be maintained in aqueous solutions during thermal cycling—conditions that are vital for proper DNA amplification. The Teflon coating serves a dual purpose in this regard: firstly, it helps keep the two glass surfaces slightly separated, allowing reaction chambers of about 20 μm in height to form between them; secondly, it helps prevent the nail varnish from entering the reaction chambers when the varnish is being applied. This is important, for any leakage of nail varnish into a reaction chamber can compromise the results in that chamber. Even if an advanced thermal cycler with humidification is being used, use of the Teflon-coated slides is still recommended, as the hydrophobicity of the Teflon combined with the pressure applied by a coverslip helps spread small volumes of reaction cocktail over the entire sample region, without forcing much fluid out at the periphery. Glass slides should be prepared according to *Protocol 1*.

Protocol 1. Silanization of slides

Equipment and reagents

- Aminopropyltriethoxysilane (APES; Sigma A-3648); prepare a fresh 2% solution just before use by mixing 5 ml of APES and 250 ml acetone[a]
- Glass slides[b]
- Coplin jars and glass staining dishes

Method

1. Put the 2% APES solution into a Coplin jar or glass staining dish and dip the glass slides in the APES for 60 sec.
2. Dip the slides five times into another vessel filled with 1000 ml of distilled water.
3. Repeat step 2 three or four times, changing the water each time.
4. Air dry in a laminar-flow hood from a few hours to overnight, then store the slides in a sealed container at room temperature. Try to use slides within 15 days of silanization.

[a] This quantity of 2% APES is sufficient to treat 200 glass slides.
[b] Heavy, Teflon-coated glass slides with three 10, 12, or 14 mm diameter wells for cell suspensions, or single oval wells for tissue sections, are available from Cel-Line Associates, Newfield, NJ or Erie Scientific, Portsmouth, NH. The advantages of using such slides are described in the text.

2.2 Preparation of cells and tissues

2.2.1 Cell suspensions

In order to use peripheral blood leucocytes, these cells must first be isolated on a Ficoll-Hypaque density gradient. Tissue-culture cells or other single-cell suspensions can also be used. Cell suspensions should be prepared according to *Protocol 2*.

Protocol 2. Preparation of cell suspensions

Method

1. Wash the cells with 1 × PBS twice.
2. Resuspend the cells in PBS at a concentration of 2×10^6 cells/ml.
3. Add at least 10 μl of cell suspension to each well of the slide using a micropipette and spread across the surface of the slide.
4. Air dry the slide in a laminar-flow hood.

2.2.2 Routinely fixed, paraffin-embedded tissue

Routinely fixed paraffin-embedded tissue sections can be amplified quite successfully. This permits the evaluation of individual cells in the tissue for the presence of a specific RNA or DNA sequence. For this purpose, tissue sections are placed on specially designed slides that have single wells (see *Protocol 3*). In our laboratory, we routinely use placental tissues, central nervous system (CNS) tissues, cardiac tissues, etc., which are sliced to a 3–5 μm thickness. Other laboratories prefer to use sections up to 10 μm in thickness, but in our experience, amplification is often less successful with the thicker sections, and multiple cell layers can lead to difficulty in interpretation due to superimposition of cells. However, if tissues that contain particularly large cells—such as ovarian follicles—are being used then thicker sections may be appropriate.

Protocol 3. Preparation of tissue sections

Equipment and reagents

- Xylene and 100% ethanol: electron microscopy (EM) grade (benzene-free)
- Teflon-coated glass slides with single oval wells.[a]

Method

1. Place the tissue section on the glass surface of the slide.
2. Incubate the slides in an oven at 60–80 °C (depending on type of paraffin used to embed the tissue) for 1 h, to melt the paraffin.
3. Dip the slides in EM grade xylene solution for 5 min, then in EM grade 100% ethanol for 5 min. Repeat these washes two or three times, in order to rid the tissue of paraffin completely.
4. Dry the slides in an oven at 80 °C for 1 h.

[a] Available from Cel-Line Associates, Newfield, NJ or Erie Scientific, Portsmouth, NH.

2.2.3 Frozen sections

It is possible to use frozen sections for *in situ* amplification; however, the morphology of the tissue following the amplification process is generally not as good as with paraffin sections. It seems that cryogenic freezing of tissue, combined with the lack of paraffin substrate during slicing, compromises tissue integrity. Usually, thicker slices must be made, and the tissue 'chatters' in the microtome. As any clinical pathologist will relate, definitive diagnoses are made from paraffin sections, and this rule of thumb seems to extend to the amplification procedure as well.

It is very important to use tissues which were frozen in liquid nitrogen or were placed on dry ice immediately after they were harvested before autolysis began to take place. If tissues were frozen slowly by placing them in –70°C then eventually ice crystals will form inside the tissues creating a gap which will distort the morphology.

To use frozen sections, use as thin a slice as possible (down to 4–6 μm), apply to the slide, dehydrate for 10 min in 100% methanol (except when the surface antigens are lipoprotein, as these will be denatured in methanol; in such cases use 2% paraformaldehyde or other reagent), and air dry in a laminar-flow hood. Then, proceed with heat treatment as described in Section 3.

3. *In situ* amplification: DNA and RNA targets

3.1 Basic preparation (all protocols)

For all sample types, *Protocol 4* gives the basic preparatory work which must be done before any amplification–hybridization procedure. The flow chart is depicted in *Figure 1.*

Protocol 4. Basic preparation

Equipment and reagents

- 10 × phosphate-buffered saline (PBS) stock solution, pH 7.2–7.4: dissolve 20.5 g $NaH_2PO_4 \cdot H_2O$ and 179.9 g $Na_2HPO_4 \cdot 7H_2O$ (or 95.5 g Na_2HPO_4) in approximately 4 litres of double-distilled water. Adjust to the required pH (7.2–7.4). Add 701.3 g NaCl and make up to a total volume of 8 litres.
- Proteinase K (Sigma) stock solution: dilute the solid to 1 mg/ml in distilled water; aliquot and store at –20°C; to make a working solution (6 μg/ml), dilute 1 ml of the stock solution in 150 ml of 1 × PBS
- 4% paraformaldehyde: add 24 g paraformaldehyde (Merck Ultrapure, cat. no. 4005) to 600 ml of 1 × PBS, heat at 65°C for 10 min, add four drops of 10 M NaOH when the solution starts to clear, stir, and adjust to neutral pH. Cool to room temperature and filter through Whatman no. 1 filter paper.

A. *Heat treatment*

1. Place the slides with adherent tissue or cells on a heat block at 105°C for 5–30 sec to stabilize the cells or tissue on the glass surface of the slide.[a]

Protocol 4. *Continued*

B. *Fixation and washes*

1. Place the slides in a solution of 4% paraformaldehyde in PBS (pH 7.4) for 4 h at room temperature. Use of Coplin jars or staining dishes facilitates these steps.
2. Wash the slides once with 3 × PBS for 10 min, agitating periodically with an up and down motion.
3. Repeat step 2 twice using 1 × PBS, using fresh PBS for the second wash.
4. At this point,[b] slides with adherent tissue can be stored at –80 °C until use. Before storage, dehydrate with 100% ethanol.

C. *Proteinase K treatment (see Section 3.1.1)*

1. Treat the samples with proteinase K working solution in PBS for 5–60 min at room temperature or at 55 °C.
2. After 5 min, look at the cells under the microscope at 400×. If the majority of the cells of interest exhibit uniform, small, round 'salt-and pepper' dots on the cytoplasmic membrane, then stop the treatment immediately with Step 3. Otherwise, continue treatment for another 5 min and re-examine.
3. After proper digestion, heat the slides on a block at 95 °C for 2 min to inactive the proteinase K.
4. Rinse the slides in 1 × PBS for 10 sec.
5. Rinse the slides in distilled water for 10 sec.
6. Air dry.

[a] This step is absolutely critical, and experimentation with different periods may be required in order to optimize the heat treatment for specific tissues. Our laboratory routinely uses 90 sec for DNA target sequences and 5–10 sec for RNA sequences. The shorter incubation is recommended for RNA targets, because certain mRNAs may be unstable at high temperature.
[b] If biotinylated probes or peroxidase-based colour development are to be used, the samples should further be treated with a 0.3% solution of hydrogen peroxide in 1 × PBS, in order to inactivate any endogenous peroxidase activity. Incubate the slides overnight, either at 37 °C or at room temperature. Then wash the slides once with 1 × PBS.
If other probes are to be used, proceed directly to the following proteinase K digestion, which is perhaps the most critical step in the protocol.

3.1.1 Optimization of digestion

The time and temperature of incubation should be optimized carefully for each cell line or tissue-section type. If there is too little digestion, the cytoplasmic and nuclear membranes will not be sufficiently permeable to primers and enzyme, and amplification will be inconsistent. On the other hand, if there is too much digestion, the membranes will either deteriorate during

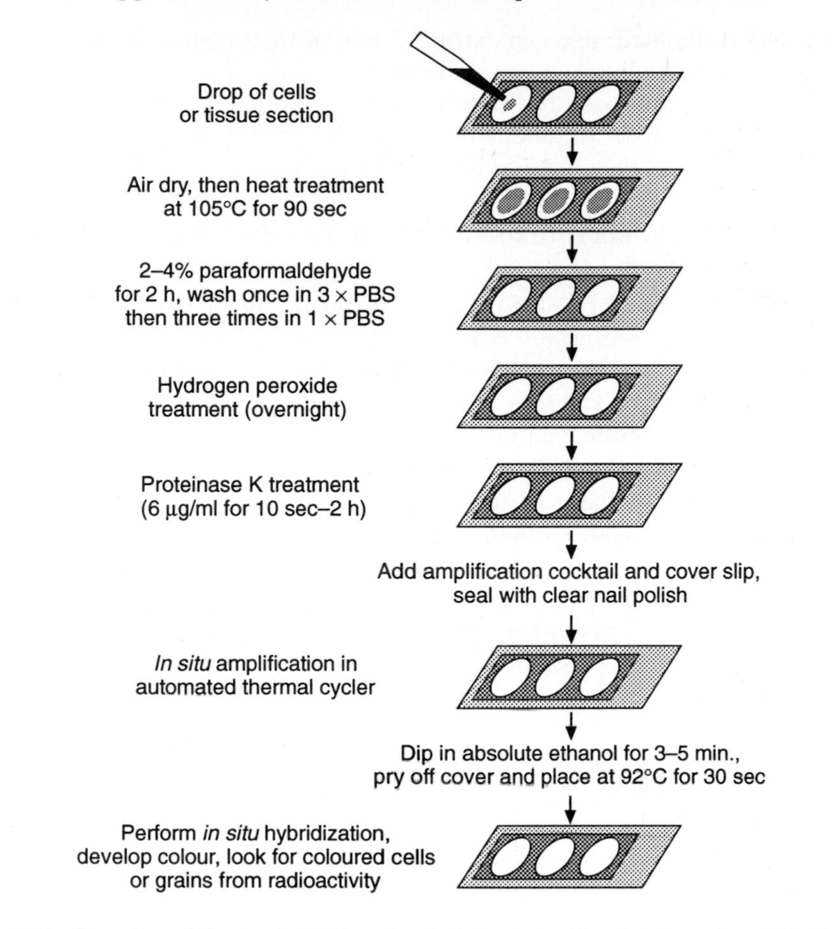

Figure 1. Overview of the *in situ* PCR protocol. Refer to text for details and modifications.

repeated denaturation or signals will leak out. In the former case cells will not contain the signal and high background will result. In the latter case, many cells will show pericytoplasmic staining, representing the leaked signals passing into cells not containing true signal. Attention to detail here can often mean the difference between success and failure, and this procedure should be practised rigorously with extra sections before continuing on to the amplification steps.

In our laboratory, proper digestion parameters vary considerably with tissue type. Typically, lymphocytes will require 5–10 min at 25 °C or room temperature, CNS tissue will require 12–18 min at room temperature, and paraffin-fixed tissue will require 15–30 min at room temperature. However, these periods can vary widely and the appearance of the 'salt and pepper dots' is the important factor. Unfortunately, the appearance of these dots is less prominent in paraffin sections. The critical importance of these dots should not be underestimated, since an extra 2–3 min of treatment after the appearance of dots will result in leakage of signals.

An alternative approach to the 'observation of dots' method is to select a constant time and treat the slides in varying concentrations of proteinase K. For example slides can be treated for 15 min in 1–6 μg/ml of proteinase K.

3.2 Reverse transcriptase variation: *in situ* RNA amplification

There are two choices for detecting an RNA signal. The first and more elegant method is simply to use primer pairs that flank spliced sequences of mRNA, as these particular sequences will be found only in RNA and will be split into sections in the DNA. Thus, by using RNA-specific primers, the following DNase step can be omitted, allowing direct reverse transcription. The second, more brutal, yet often necessary approach is to treat the cells or tissue with DNase after the proteinase K digestion. This step destroys all of the endogenous DNA in the cells so that only RNA survives to provide signals for amplification.

Note that all reagents for RT *in situ* amplification should be prepared with RNase-free water (i.e. water treated with diethylpyrocarbonate (DEPC)). In addition, the silanized glass slides and all glassware should be RNase-free, which can be achieved by baking the glassware overnight in an oven before use in the RT amplification procedure.

Next, DNA copies of the targeted RNA sequence must be made so that the signal can be amplified. *Protocol 5* gives typical cocktails for the reverse transcription reaction.

Protocol 5. Reverse transcriptase *in situ* amplification

Equipment and reagents

- RNase-free DNase, such as 1000 U/μl Boehringer RQ1 DNase, cat no. 776785
- DNase buffer: 40 mM Tris–HCl, pH 7.4, 6 mM $MgCl_2$, 2 mM $CaCl_2$
- 10× reaction buffer (for AMV or MMLV RT): 250 mM Tris pH 8.3, 375 mM KCl, 15 mM $MgCl_2$
- AMV or MMLV RT
- dATP, dCTP, dGTP, dTTP (each 10 mM)

A. *DNase treatment*[a]

1. Prepare an RNase-free DNase solution comprising 100 U/μl DNase in DNase buffer.
2. Add 10–15 μl of solution to each well.
3. Incubate the slides at 37 °C in a humidified chamber for 1 h. If using liver tissue, this incubation should be extended for an additional hour.
4. After incubation, rinse the slides with a similar solution that was prepared without the DNase.
5. Wash the slides twice with DEPC-treated water.

B. *Reverse transcriptase reaction (see Section 3.2.1)*

1. Make one of the following cocktails:
   (a) If using AMV or MMLV RT (see Section 3.2.1):
   - 10 × reaction buffer 4.0 μl
   - 10 mM dATP 2.0 μl
   - 10 mM dCTP 2.0 μl
   - 10 mM dGTP 2.0 μl
   - 10 mM dTTP 2.0 μl
   - RNasin at 40 U/μl (Promega)[b] 0.5 μl
   - 20 μM downstream primer 1.0 μl
   - AMV RT (20 U/μl) 0.5 μl
   - DEPC-treated water 6.0 μl
   - (Total volume 20.0 μl)

   (b) If using Superscript II enzyme (see Section 3.2.1) from BRL:
   - 5 × reaction buffer[c] 4.0 μl
   - 10 mM dATP 2.0 μl
   - 10 mM dCTP 2.0 μl
   - 10 mM dGTP 2.0 μl
   - 10 mM dTTP 2.0 μl
   - RNasin at 40 U/μl (Promega)[b] 0.5 μl
   - 20 μM downstream primer 1.0 μl
   - Superscript II (200 U/μl) 0.5 μl
   - 0.1 M DTT 2.0 μl
   - DEPC-treated water 4.0 μl
   - (Total volume 20.0 μl)
2. Add 10–15 μl of either cocktail to each well. Carefully place the coverslip on top of the slide.
3. Incubate at 42 °C or 37 °C for 1 h in a humidified atmosphere.
4. Incubate the slides at 92 °C for 2 min.
5. Remove the coverslip and wash twice with distilled water. Proceed with the amplification procedure, which is the same for both DNA- and RNA-based protocols.

[a] For cells that are particularly rich in RNase, add 1000 U/ml placental RNase inhibitor (e.g. RNAsin) plus 1 mM DTT to the DNase solution. Some investigators prefer to use a long incubation period with a lower concentration of DNase (1 U/ml for 18 h).
[b] RNAsin inhibits ribosomal RNases. These should be used for optimal yields.
[c] As supplied with enzyme.

3.2.1 Reverse transcriptases

Avian myeloblastosis virus (AMV) RT and Moloney murine leukaemia virus (MMLV) RT give comparable results in our laboratory. Other RTs will probably work also. However, it is important to read the manufacturer's descriptions of the RT and to make certain that the proper buffer is used.

An alternative RT is available which lacks RNase activity. Called 'Superscript II', it is available from BRL Lifesciences, and is suitable for reverse transcription of long mRNAs. It is also suitable for routine RT amplification, and in our laboratory has proven to be more efficient than the two enzymes described above.

3.2.2 Primers and target sequences

In our laboratory, we simply use anti-sense downstream primers for our gene of interest, as we already know the sequence of most genes we study. However, oligo(dT) primers can be used as an alternative first to convert all mRNA populations into cDNA, and then to perform the *in situ* amplification for a specific cDNA. This technique may be useful when amplification of several different gene transcripts is being performed at the same time in a single cell. For example, if the detection of expression of various cytokines is required, one can use an oligo(dT) primer to reverse transcribe all of the mRNA copies in a cell or tissue section. Then, more than one type of cytokine can be amplified and the individual types detected with different probes which develop into different colours (see Section 4.2).

In all reverse transcription reactions, it is advantageous to reverse transcribe only relatively small fragments of mRNA (<1500 bp). Larger fragments may not be completely reverse-transcribed due to the presence of secondary structures. Furthermore, the RTs—at least AMV and MMLV RTs—are not very efficient in transcribing large mRNA fragments. However, this size restriction does not apply to amplification reactions that are exclusively DNA, for the polymerase enzyme copies nucleotides better. In *in situ* DNA reactions, we routinely amplify genes up to 300 bp. The following points should be kept in mind:

- both sense and anti-sense primers should be 14–22 bp long
- primers should contain GG, CC, GC, or CG base pairs at the 3′-ends to facilitate complementary strand formation
- the preferred GC content of the primers is 45–55%
- try to design primers that do not form intra- or interstrand base pairs
- the 3′-ends of the primers should not be complementary to each other, or they will form primer dimers
- a reverse transcription primer can be designed so that it does not contain secondary structures

3.2.3 Annealing temperature

Annealing temperatures for reverse transcription and for DNA amplification can be chosen according to the following formula:

$$T_m \text{ of the primers} = 81.5\,^{\circ}\text{C} + 16.6(\log M) + 0.41(\text{G} + \text{C}\%) - 500/n$$

where n is the length of primers and M is the molarity of the salt in the buffer, usually 0.047 M for DNA reactions and 0.070 M for reverse transcription reactions (see below).

If using AMV RT, the value will be lower according to the following formula:

$$T_m \text{ of the primers} = 62.3\,°\text{C} + 0.41(\text{G} + \text{C}\%) - 500/n$$

Optimization of the annealing temperature should be carried out first with solution-based reactions. It is important to know the optimal temperature before attempting to conduct *in situ* amplification, as *in situ* reactions are simply not as robust as solution-based ones. This may be because primers do not have easy access to DNA templates inside cells and tissues, as numerous membranes, folds, and other small structures can prevent primers from binding homologous sites as readily as they do in solution-based reactions.

There are two other ways of determining true annealing temperatures:

- using 'Robocycler', a thermocycler recently developed by Stratagene (La Jolla, CA) and designed for determination of actual annealing temperature
- using so-called 'Touchdown' PCR (22)

The logic of determining the correct annealing temperature for *in situ* PCR is that, during amplification, spurious products often appear in addition to those desired. Therefore, even if the cells do not contain DNA homologous to the primer sequences, many artefactual bands may appear. Many protocols have appeared in the literature to overcome false priming, including hot-start, use of dimethylsulfoxide, formamide and anti-*Taq*-polymerase antibodies (i.e. Clontech's *Taq*Start). It is important to remember that false priming will occur if the annealing temperature does not match the melting temperature. Thus, annealing temperatures significantly above the T_m will yield no products and temperatures too far below the T_m often will give unwanted products due to false priming. Therefore determination of optimal annealing temperature is extremely important.

Usually, primer annealing is optimal at 2°C above T_m. However, this formula provides only an approximate temperature for annealing, since base-stacking, near-neighbour effect, and buffering capacity may play a significant role for a particular primer.

Recently, MJ Research (Watertown, MA) have devised a thermocycler which can perform *in situ* gene amplification both on slides and in solution (tubes) simultaneously, in the same block. That kind of thermocycler can be very useful in determining the optimal amplification of a gene of interest.

3.2.4 The 'hot-start' technique

There is much debate as to whether a 'hot start' helps to improve the specificity and sensitivity of amplification reactions. In our laboratory, we find the 'hot start' adds no advantage in this regard; rather, it adds only technical difficulty to the practice of the *in situ* technique. However, a variation of the 'hot start' has recently been reported. In this procedure, PCR cocktail (containing *Taq* polymerase) contains an anti-*Taq*-polymerase antibody, which keeps the *Taq* enzyme in the cocktail 'blocked' until the first denaturation step when the anti-*Taq*-polymerase antibody is denatured and restores full *Taq* polymerase activity. This modification essentially serves the same function as the 'hot start' procedure but without its difficulties.

Protocol 6. Amplification protocol (all types)

Equipment and reagents

- Forward and reverse primers
- A stock solution containing 10 mM of each of dATP, dCTP, dGTP and dTTP
- *Taq* DNA polymerase e.g. AmpliTaq 5 U/μl[a]
- 10 × PCR buffer[b]
- Clear nail varnish
- 2 × SSC: 0.3 M NaCl, 0.03 M sodium citrate, pH 7.0

Method

1. Prepare the following amplification cocktail:

• 25 μM forward primer (e.g. SK38 for HIV-1)	5.0 μl
• 25 μM reverse primer (e.g. SK39 for HIV-1)	5.0 μl
• dNTP stock solution	2.5 μl
• *Taq* polymerase (AmpliTaq 5 U/μl)	1.0 μl
• 10 × PCR buffer	10.0 μl
• H_2O	76.5 μl
• (Total volume	100.0 μl)

2. Layer 10–15 μl of amplification solution on to each well with a micropipette so that the whole surface of the well is covered with the solution. Be careful not to touch the surface of the slide with the tip of the pipette.
3. Add a glass coverslip (22 mm × 60 mm) and carefully seal the edge of the coverslip to the slide with two coats of clear nail varnish. If using tissue sections, use a second slide instead of a coverslip.[c]
4. Allow nail varnish to dry completely.
5. Place the slides in a thermocycler.
6. Run 30 cycles of the following amplification protocol:[d]
   - 94°C, 30 sec
   - optimal annealing temperature (see Section 3.2.3), e.g. 45°C, 1 min
   - 72°C, 1 min
7. When thermal cycling is complete, dip slides in 100% ethanol for at least 5 min, in order to dissolve the nail varnish. Prise off the coverslip using a razor blade or other fine blade: the coverslip generally pops off quite easily. Scratch off any remaining nail varnish on the outer edges

of the slide so that fresh coverslips will lie evenly in the subsequent hybridization/detection steps.

8. Place the slides on a heat block at 92°C for 1 min: this treatment helps immobilize the intracellular signals.
9. Wash the slides with 2 × SSC at room temperature for 5 min.

[a] Other thermostable polymerases have also been used quite successfully
[b] This is usually supplied with the thermostable polymerase. A suitable formula is: 100 mM Tris–HCl (pH 8.3), 500 mM KCl, 15 mM $MgCl_2$, 0.01% gelatin.
[c] Be certain to paint the nail varnish carefully all around the edge of the coverslip or the edges of the dual slide, as the nail varnish must completely seal the coverslip–slide assembly in order to form a small 'reaction chamber' that can contain the water vapour during thermal cycling. For effective sealing, do not use coloured nail varnish or any especially 'runny'nail varnish. Proper sealing is very important, for this keeps reaction concentrations consistent through the thermal cycling procedure, and concentrations are critical to proper amplification. Be certain to apply the nail varnish very carefully so that none of the varnish gets into the actual chamber where the cells or tissues reside. If any nail varnish does enter the chamber, discard that slide for the results will be questionable. The painting of nail polish is truly a learned skill, so it is strongly recommended that researchers practise this procedure several times with mock slides before attempting an experiment. In the case of thick tissue sections, it is best to use another identical blank slide for the cover instead of a coverslip. Apply the amplification cocktail to the appropriate well of the blank slide, place an inverted tissue-containing slide on top of the blank slide, and seal the edges as described. Invert the slide once again so that the tissue-containing slide is on the bottom. This technique can be modified to accommodate a hot start (see Section 3.2.4).
[d] These times/temperatures will probably require optimization for the specific thermocycler being used. Furthermore, the annealing temperature should be optimized, as described in Section 3.2.3. These particular incubation parameters work well with SK38 and SK39 primers for the HIV-1 *gag* sequence, when amplified in an MJ Research PTC-100-60 or PTC-100-16MS thermal cycler.

3.2.5 Thermal cyclers

Various thermocycler technologies will work in this application; however, some instruments work much better than others. In our laboratory, we use two types: a standard, block-type thermocycler that normally holds 60 0.5 ml tubes but which can be adapted with aluminium foil, paper towels, and a weight to hold four to six slides. We also use dedicated thermocyclers that are specifically designed to hold 12 or 16 slides. We understand that other laboratories have used stirred-air, oven-type thermocyclers quite successfully; however, we have also heard that there are sometimes problems with the cracking of glass slides during cycling. Thermocyclers dedicated to glass slides are now available from several vendors, including Barnstead Thermolyne, Coy Corporation, Hybaid, and MJ Research. Our laboratory has used an MJ Research PTC-100-16MS and a PTC-100-16MS quite successfully. Recently, this company has combined the slide and tubes into a single block, allowing the simultaneous confirmation of *in situ* amplification in a tube. Furthermore, there are newer designs of thermal cycler which incorporate humidification chambers. We do not yet have sufficient experience with this technology to verify whether such machines can eliminate the need for sealing the slides with nail varnish during thermal cycling. Nonetheless, the humidified instruments are especially useful in the reverse transcription and hybridization steps, where otherwise a humidified incubator is needed.

We suggest that you follow the manufacturer's instructions on the use of your own thermocycler, bearing in mind the following points:

(a) Glass does not easily make good thermal contact with the surface on which it rests. Therefore, a weight to press down the slides and/or a thin layer of mineral oil to fill in the interstices will help thermal conduction. If using mineral oil, make certain that the oil is well smeared over the glass surface so that the slide is not merely floating on air bubbles beneath it.

(b) The top surfaces of slides lose heat quite rapidly through radiation and convection; therefore, use a thermocycler that envelops the slide in an enclosed chamber (as in some dedicated instruments), or insulate the tops of the slides in some manner. Insulation is particularly critical when using a weight on top of the slides, for the weight can serve as an unwanted heat sink if it is in direct contact with the slides.

(c) Good thermal uniformity is imperative for good results—poor uniformity or irregular thermal change can result in cracked slides, uneven amplification, or completely failed reactions. If adapting a thermocycler that normally holds plastic tubes, use a layer of aluminium foil to spread out the heat.

3.2.6 Direct incorporation of non-radioactive labelled nucleotides

Several non-radioactively labelled nucleotides are available from various sources (e.g. biotin-dCTP, digoxigenin-11-dUTP, etc.). These nucleotides can be used to label amplification products directly with subsequent detection of the directly labelled *in situ* amplification products using secondary agents and chromogens. However, in our opinion, as well as in the opinion of several other laboratory groups, the greatest specificity is only achieved by conducting amplification followed by subsequent *in situ* hybridization (ISH) (1–16). In the direct labelling protocols, non-specific incorporation can be significant, and even if this incorporation is minor, it still leads to false-positive signals similar to non-specific bands in gel electrophoresis following solution-based DNA- or RT-amplification. Therefore, we strongly discourage the direct incorporation of labelled nucleotides as part of an *in situ* amplification protocol.

The only exception to this recommendation is for screening a large number of primer pairs for optimization of a specific assay. To achieve this, add to the amplification cocktail detailed in *Protocol 6* 4.3 μM labelled nucleotide (14-biotin-dCTP or 14-biotin-dATP or 11-digoxigenin-dUTP) along with unlabelled nucleotide to achieve a 0.14 mM final concentration. Also, if the perfect annealing system has been worked out, using either Robocycler or an equivalent system, then direct incorporation can be used without fear of non-specific labelling, which we have discussed elsewhere in detail (8).

4. Special application of *in situ* amplification

4.1 *In situ* amplification and immunohistochemistry

Immunohistochemistry and *in situ* amplification can be performed simultaneously in a single cell. For this purpose, we first fix cells or frozen sections of tissue, which are already placed on slides, with 100% methanol for 10 min. Then, slides are washed in PBS. After that labelling of surface antigen(s) can be carried out by standard immunohistochemistry methods (e.g. incubate fluorescein isothiocyanate (FITC)-conjugated antibody at 37 °C for 1 h, wash, and then fix cells or tissue section in 4% paraformaldehyde for 2 h). In various pathology laboratories, it has been noted that many specific surface antigens can withstand 10% formalin and other routine histopathology procedures and will still bind specific monoclonal antibodies. If any of these immunohistochemistry panels is being used then routinely prepared paraffin sections can be used for the detection of cellular antigens. Then, the tissue is prepared for *in situ* amplification as described earlier.

Following development of the colour of the amplified product in the post-hybridization step, the cells or tissue can be viewed under UV/visible light in an alternating fashion, to detect two signals in a single cell.

4.2 Multiple signals, multiple labels in individual cells

DNA, mRNA, and protein can all be detected simultaneously in individual cells. As described in the above section, proteins can be labelled using FITC-labelled antibodies. Then, both RNA and DNA *in situ* amplification can be performed in the cells and, if primers for spliced mRNA are used and if these primers do not bind any DNA sequences, both DNA and RT amplification can be carried out simultaneously. The reverse transcription step still needs to be performed but this time without pre-DNase treatment. Subsequently, products can be labelled with different kinds of probes, resulting in different colours of signal. For example, proteins can be labelled with FITC, mRNA with a rhodamine-conjugated probe (20 colours are available), and DNA with a biotin–peroxidase probe. Each will show a different signal within an individual cell.

5. Hybridization

5.1 Summary of the basic concepts of *in situ* hybridization (see also Chapter 3)

In this section some of the pertinent aspects of ISH are highlighted. This technique is a very sensitive method for detecting specific nucleic acid sequences inside cells. The method utilizes the specific annealing of a labelled nucleic acid probe to complementary sequences within intact fixed cells or chromosomes prepared and adhered to the surface of a microscope slide. This is followed by visualization, through a microscope, confocal or image analysis-based system, of the probe location in the specimen. Therefore, ISH involves annealing of a complementary probe to a RNA or DNA gene sequence *in situ* and detection of the probe by some visible method. The technique is essentially the same as Southem blotting except that hybridization is taking place *in situ* (inside the cells).

The power of ISH lies in the ability to determine the presence and localization of specific nucleic acids within an intact cell. It uniquely provides localization of nucleic acids superimposed on the cellular and sub-cellular detail of the specimen, which cannot be detected by Southern blots, Northern blots, and dot blot filter hybridization assays.

It is important to realize that, unlike hybridization of nucleic acids in solution or immobilized on a solid support (e.g. nylon membrane), the nucleic acids detected *in situ* are in a complex matrix and gene sequences to be hybridized are not readily accessible. Hence, many parameters for ISH are empirically determined.

The ISH technique has been successfully applied in both the research and clinical settings. However, one single, easy-to-use universal procedure has not been developed. Therefore, specific needs of the diagnostic or research goals must be considered in choosing a suitable protocol.

In comparing ISH with other methods it is essential to appreciate what is being detected. For example, irnmunocytochemical methods localize protein within a cell or on the cell surface and therefore identify gene expression. However, these assays cannot yield useful information on post-translational processing of the gene product or differentiate between the uptake and storage of the protein and the site of synthesis of the protein. In addition, several hundred copies of the proteins are required to be able to identify an expression signal. mRNA extraction methods utilize the isolation of nucleic acids from cells (filter hybridization assays), and therefore can dilute the target found in only a few cells by many cells with little or no target. These methods provide no information on distribution: they give only an average measurement of the nucleic acid target present in the mixed cell population. Therefore ISH is a very powerful technique when the target is focally distributed within a single cell or certain histological cell type within a tissue. Consequently, ISH is more sensitive than filter assays if the gene expression is taking place in a small sub-population of cells. The major limitation of ISH is its relative insensitivity by comparison with *in situ* PCR. Using ISH, as few as 20 copies of mRNA can be detected. However, that degree of sensitivity can only be achieved in a few highly specialized laboratories and for a limited number of specific genes. More realistically, detection of >80 copies/cell would be an achievable goal for a laboratory not specialized in ISH. In order to detect single-copy genes or to detect very low level gene expression (where there are only a few copies of the mRNA of interest), the gene sequences can be amplified *in situ* by DNA or RNA (RT) *in situ* PCR followed by ISH to detect the amplicons.

Analysing gene expression by ISH after reverse transcription can provide information on the site of mRNA synthesis which provides information about the cellular origin of protein synthesis and demonstrates the amount of synthesis (level of gene expression). This permits an understanding of the cell types involved in certain protein synthesis and in certain gene regulation, and of the cell types infected by various viral or other infectious agents. In addition, by combining immunohistochemistry, differential expression of a gene in different cell types or different stages of development can be analysed at the microscopic level.

5.1.1 Choice of probes for ISH

Many ISH protocols employ ^{3}H- or ^{35}S-labelled nucleic acid probes, followed by autoradiographic detection. Although this method can be very sensitive and although ^{3}H-labelled probes generate well resolved autoradiographic signals, it is time consuming and technically difficult. Other isotopes emitting high levels of radiation can be used but give non-specific background.

Non-isotopic methods for ISH offer the advantages of probe stability, sensitivity, spatial resolution, and speed. The non-isotopic adaptations are generally simpler and faster than autoradiography, and the sensitivity of non-radioactive methods has increased over the years as the parameters influencing hybridization efficiency and signal specificity have been optimized. Factors contributing to increased use of non-isotopic methods include:

- faster colour development
- chemically stable probes with no special disposal requirements
- availability of different labelling and detection systems which can be used to facilitate the analysis of several probes simultaneously

5.1.2 *In situ* hybridization

This is detailed in *Protocols 7* and *8*.

Protocol 7. *In situ* hybridization

Equipment and reagents

- Probe
- Deionized formamide
- 20 × SSC: dissolve 175.3 g of NaCl and 88.2 g of sodium citrate in 800 ml of water. Adjust the pH to 7.0 with a few drops of 10 M NaOH. Adjust the volume to 1 litre with water. Sterilize by autoclaving.
- Denhardt's solution
- Sonicated salmon sperm DNA, denatured by heating at 94°C for 10 min before adding to the hybridization buffer
- 10% SDS

Protocol 7. *Continued*

Method

1. Prepare the following solution:[a,b]
   - Probe (biotinylated, or digoxigenin) 2 μl
   - Deionized formamide 50 μl
   - 20 × SSC 10 μl
   - 50 × Denhardt's solution 20 μl
   - 10 mg/ml sonicated salmon sperm DNA 10 μl
   - 10% SDS 1 μl
   - H_2O 7 μl
   - Total volume 100 μl
2. Add 10 μl of hybridization mixture to each well and add coverslips.
3. Heat slides on a block at 95°C for 5 min to denature the double-stranded DNA.
4. Incubate slides at 48°C for 2–4 h in a humidified atmosphere.[c]

[a] The final concentrations are 20–50 pmol/μl of the appropriate probe, 50% deionized formamide, 2 × SSC buffer, 10 × Denhardt's solution, 0.1% sonicated salmon sperm DNA, and 0.1% SDS. Hybridization mix without probe can be made up in larger volumes.
[b] 2% BSA can be added if non-specific binding is a problem. Add 10 μl of 20% BSA and reduce the amount of water.
[c] Optimal hybridization temperature is a function of the T_m of the probe. This must be calculated for each probe, as described earlier. However, the hybridization temperatures used should not be too high. If that circumstance occurs, then the formula for the hybridization solution should be modified and instead of 50% formamide, 40% formamide should be substituted (described further in the *in situ* hybridization section of reference 13.)

Protocol 8. Signal development

Equipment and reagents

- PBS
- Streptavidin–peroxidase conjugate (Sigma) dissolved in PBS to make a stock of 1 mg/ml; just before use, dilute the stock solution in sterile PBS in a 1:30 ratio
- 50 mM acetate buffer, pH 5.0: add 74 ml of 0.2 N acetic acid (11.55 ml glacial acid/litre) and 176 ml of 0.2 M sodium acetate (27.2 g sodium acetate trihydrate in 1 litre) to 1 litre of deionized water and mix.
- 3′-Amino-9-ethylcarbazole (AEC): dissolve one AEC tablet (Sigma) in 2.5 ml of *N,N*-dimethylformamide. Store at 4°C in the dark.
- 30% (v/v) hydrogen peroxidase
- Blocking solution: 50 mg/ml BSA in 100 mM Tris–HCl (pH 7.8), 150 mM NaCl, and 0.2 mg/ml sodium azide
- Conjugate dilution buffer: 100 mM Tris–HCl, 150 mM $MgCl_2$, 10 mg/ml BSA and 0.2 mg/ml sodium azide
- Buffer A: 100 mM Tris–HCl, pH 7.5, 150 mM NaCl
- Alkaline substrate buffer: 100 mM Tris–HCl (pH 9.5), 150 mM NaCl, and 50 mM $MgCl_2$
- Nitroblue Tetrazolium (NBT), 75 mg/ml in 70% (v/v) dimethylformamide
- 5-Bromo-4-chloro-3-indolylphosphate (BCIP), 50 mg/ml in 100% dimethylformamide

A. *Peroxidase-based colour development*

1. Wash the slides in 1 × PBS twice for 5 min each time.

2. Add 10–15 μl of streptavidin–peroxidase complex and gently apply the coverslips.
3. Incubate the slides at 37 °C for 1 h.
4. Remove the coverslip and wash the slides with 1 × PBS twice for 5 min each time.
5. Mix the following chromogen solution and add 100 μl to each well:

• 50 mM acetate buffer, pH 5.0	5 ml
• 30% H_2O_2	25 μl
• AEC	250 μl

6. Incubate the slides at 37 °C for 10 min in the dark to develop the colour, then observe the slides under a microscope. If the colour is not strong, develop for another 10 min.
7. Rinse the slides with tap water and allow to dry.
8. Add one drop of 50% glycerol in PBS and apply the coverslips.
9. Examine with an optical microscope; positive cells will be stained a brownish red.

B. *Alkaline phosphatase-based colour development*

1. After hybridization, remove the coverslip and wash the slides with two changes of 2 × SSC at room temperature for 15 min each.
2. Cover each well with 100 μl of blocking solution and place the slides flat in a humid chamber at room temperature for 15 min.
3. Prepare a working conjugate solution by mixing 10 μl of streptavidin–alkaline phosphatase conjugate (40 μg/ml stock) with 90 μl of conjugate dilution buffer for each well.
4. Remove the blocking solution from each slide by touching the edge of the slide with absorbent paper.
5. Cover each well with 100 μl of working conjugate solution and incubate in the humid chamber at room temperature for 15 min. Do not allow the tissue to dry after adding the conjugate.
6. Wash the slides by soaking in two changes of buffer A for 15 min each at room temperature.
7. Wash the slides once in alkaline substrate buffer (see below) at room temperature for 5 min.
8. Pre-warm 50 ml of alkaline substrate buffer to 37 °C in a Coplin jar. Just before adding the slides, add 200 μl of NBT solution and 166 μl of BCIP solution and mix well.
9. Incubate the slides in the NBT/BCIP solution at 37 °C until the desired level of signal is achieved (usually from 10 min to 2 h). Check the

Protocol 8. *Continued*

colour development periodically by removing a slide from the NBT/BCIP solution. Be careful not to allow the tissue to dry.

10. Stop the colour development by rinsing the slides in several changes of deionized water. The tissue may now be counter-stained.

C. *Counter-staining and mounting*

1. Stain for five min at room temperature (if you are using a red indicator (like AEC) then use Gill's haematoxylin (Sigma) and if using NBT/BCIP as indicator then use Nuclear Fast Red stain as counter-stain).
2. Rinse in several changes of tap water.
3. Air dry at room temperature.
4. For permanent mounting, a water-based medium such as Crystal-Mount or GelMount can be used.
5. Apply one drop of mounting medium per 22 mm coverslip.
6. The slides may be viewed immediately if you are careful not to disrupt the coverslip. The mounting medium will dry overnight at room temperature.

6. Validation and controls

The validity of *in situ* amplification–hybridization should be examined in every run. Attention here is especially necessary in laboratories first using the technique, because occasional technical pitfalls lie on the path to mastery. In an experienced laboratory, it is still necessary to continuously validate the procedure and to confirm the efficiency of amplification. To do this, we routinely run two or three sets of experiments in multi-welled slides simultaneously, for we must not only validate amplification, but also confirm the subsequent hybridization/detection steps.

In our laboratory, we frequently work with HIV. A common validation procedure we will conduct is to mix HIV-1-infected cells and HIV-1-uninfected cells in a known proportion (e.g. 1:10, 1:100, etc.), then confirm that the results are appropriately proportionate. To examine the efficiency of amplification, we use a cell line which carries a single copy or two copies of cloned HIV-1 virus, then look to see that proper amplification and hybridization have occurred.

In all amplification procedures, we use one slide as a control for non-specific binding of the probe. Here we hybridize the amplified cells with an unrelated probe. We also use HLA-DQα probes and primers with human PBMCs as positive controls, to check various parameters of our system.

If using tissue sections, a cell suspension lacking the gene of interest can be used as a control. These cells can be added on top of the tissue section and then retrieved after the amplification procedure. The cell suspension can then be analysed with the specific probe to see if the signal from the tissue leaked out and entered the cells floating above.

We suggest that researchers carefully design and employ appropriate positive and negative controls for their specific experiments. In the case of RT-*in situ* amplification, β-actin, β-globin, HLA-DQα, and other endogenous abundant RNAs can be used as positive markers. Of course, a RT-negative control for RT *in situ* amplification, as well as DNase and non-DNase controls, should also be used. Controls without *Taq* polymerase with primers and without primers should always be included.

References

1. Bagasra, O. (1990). *Amplifications*, March 20.
2. Bagasra, O., Hauptman, S.P., Lischner, H.W., Sachs, M., and Pomerantz, R.J. (1992). *New Engl. J. Med.*, **326**, 1385.
3. Bagasra, O., Seshamma, T., and Pomerantz, R.J. (1993). *J. Immunol. Methods*, **158**, 131.
4. Bagasra, O. and Pomerantz, R.J. (1993). *AIDS Res. Human Retroviruses*, **9**, 69.
5. Bagasra, O., Seshamma, T., Oakes, J., and Pomerantz, R.J. (1993). *AIDS*, **7**, 82.
6. Bagasra, O., Seshamma, T., Oakes, J., and Pomerantz, R.J. (1993). *AIDS*, **7**, 1419.
7. Bagasra, O., Farzadegan, H., Seshamma, T., Oakes, J. , Saah, A., and Pomerantz, R.J. (1994). *AIDS*, **8**, 1669.
8. Bagasra, O. and Pomerantz, R.J. (1994). In *Clinics of North America* (ed. R.J. Pomerantz), pp. 351–66. W. B. Saunders, Philadelphia.
9. Bagasra, O., Hui, Z., Bobroski, L., Seshamma, T., Saikumari, P., and Pomerantz, R.J.. (1995). *Cell Vision*, **2,** 425.
10. Bagasra, O., Michaels, F., Mu, Y., Bobroski, L., Spitsin, S. V., Fu, Z. F., and Koprowski, H. (1995). *Proc. Natl Acad. Sci. USA*, **92**, 12041.
11. Bagasra, O. and Pomerantz, R.J. (1995). *PCR in Neuroscience* (ed. G. Sarkar), pp. 339–57. Academic Press, New York.
12. Bagasra, O., Seshamma, T., Pastanar, J.P., and Pomerantz, R.. (1995). *Technical Advances in AIDS Research in the Nervous System* (ed. E. Majors), pp. 251–66. Plenum Press, New York.
13. Bagasra, O., Seshamma, T., Hansen, J., and Pomerantz, R. (1995). In *Current Protocols in Molecular Biology* (ed. F. Ausubel *et al.*), section 14.8.1.
14. Bagasra, O., Lavi, U., Bobroski, L., Khalili, K., Pestaner, J.P., and Pomerantz, R.J. (1995). *AIDS*, **10,** 573.
15. Embretson, J., Zupanic, M., Beneke, T., Till, M., Wolinsky, S., Ribas, J.L., Burke, A., and Haase A.T. (1993). *Proc. Natl Acad. Sci. USA*, **90**, 357.
16. Embretson, J., Zupanic, M., Ribas, J.L., Burke, A., Racz, P., Tanner-Racz, K., and Haase, A.T. (1993). *Nature*, **62**, 359.
17. Lattime, E.C., Mastrangelo, M.J., Bagasra, O., and Berd, D. (1995) *Cancer Immunol. Immunother.*, **41,** 151.

18. Mehta, A., Maggioncalda, J., Bagasra, O., Thikkavarapu, S., Saikumari, P., Nigel, F.W., and Block, T. (1995). *Virology*, **206**, 633.
19. Nuovo, G.J., Becker, J., Margiotta, M., MacConnell, P., Comite, S., and Hochman, H. (1992). *Am. J. Surg. Pathol.*, **16**, 269.
20. Nuovo, G.J. (1994). *PCR In Situ Hybridization Protocols and Applications*, 2nd edition. Raven Press, NewYork.
21. Patterson, B.K., Till, M., Otto, P., Goolsby, C., Fortado, M.R., McBride, L.J., and Wolinsky, S.M. (1993). *Science*, **260**, 976.
22. Pereira, R.F., Halford, K.W., O'Hara, M.D., Leeper, D.B., Sokolov, B.P., Pollard, M.D., Bagasra, O., and Prockop, D.J. (1995). *Proc. Natl Acad. Sci. USA*, **92**, 4857.
23. Pestaner, J.P., Bibbo, M., Bobroski, L., Seshamma, T., and Bagasra, O. (1994). *Acta Cytol.*, **38**, 676.
24. Qureshi, M.N., Barr, C.E., Seshamma, T., Pomerantz, R.J., and Bagasra, O. (1994). *J. Infect. Dis.*, **171**, 190.
25. Winslow, B.J., Pomerantz, R.J., Bagasra, O., and Trono, D. (1993). *Virology*, **196**, 849.

8

Applications and modifications of PCR *in situ* hybridization

BRUCE K. PATTERSON

1. Introduction

1.1 General

The localization of genes in specific cells and tissues has been instrumental in many disciplines for defining disease pathogenesis. This pursuit has united scientists studying such diverse topics as oncogene expression, viral infections, and plant pathology. In its earliest incarnation, molecular biologists met with pathologists to use large subgenomic probes to detect genes *in situ*. This process relied on the intrinsic amplification of genes during transcription into mRNA to accumulate enough target to be detected.

The determination that HIV is the causative agent in AIDS (1) pushed technological advances especially in the field of viral detection. Early experiments revealed that the HIV life-cycle included a proviral latent phase in which a single copy of HIV DNA is integrated into the host cell genome (2). In order to determine the cell types infected, a sensitive method had to be devised which could detect single-copy genes (3). In the early 1990s, such methods helped determine that a high viral burden was present in patients infected with HIV (4). These determinations shifted the theory of HIV pathogenesis from indirect mechanisms such as autoimmunity, superantigens, and apoptosis (5) to more direct mechanisms of viral pathogenesis. Recent work on the kinetics of HIV has verified direct pathogenic mechanisms of HIV infection (6,7).

Polymerase chain reaction (PCR) *in situ* hybridization has expanded into other areas such as fetal diagnosis (author's unpublished data), transplantation (8), and minimal residual disease detection in oncology (author's unpublished data). For a variety of human malignancies, identification of specific chromosomal and genetic abnormalities is a critical component of accurate diagnosis. Clearly, a method by which rapid identification of specific translocations and genetic alterations could be coupled with routine immunophenotypic analysis would greatly enhance the diagnostic usefulness of flow cytometric or image analysis detection and the characterization of leukaemias and lymphomas. These approaches would be not only far faster than conventional cytogenetic analyses but also cost-effective. Most importantly, development of these approaches will allow investigators to address questions which are difficult or impossible with currently available technology. *In situ* hybridization techniques in cells are limited to detecting positive cells when they comprise at least one to five per cent of the population, and thus have limited usefulness in detecting minimum residual disease or monitoring therapy. Additionally, the ability simultaneously to correlate this information with other cellular parameters is limited. Conventional PCR or reverse transcription PCR (RT-PCR) is extremely sensitive, being able to detect as few as one positive cell in one million cells; however, quantification is labour-intensive and technically difficult, and additional characteristics of the rare positive cells in the population are not available for analysis. Development of sensitive techniques for detecting genetic aberrations which determine not only the presence of an abnormality within the population but also the percentage of cells that harbour that abnormality and the characteristics of those abnormal cells, will have a significant impact potentially leading to improved diagnostic and prognostic determinations.

The purpose of this chapter is to describe basic techniques that have already been used in a variety of different applications and to illustrate how hypothesis-driven research has guided modifications and advancements of the basic technology. The single-parameter protocols, though they may be similar to other protocols already presented, have all been optimized for subsequent multiparameter analysis.

1.2 Technical

Optimal analyses of cell constituents require control of the entire process of specimen acquisition, preparation, fixation, and storage. Though this may not be practical or useful in some applications where the study of archival cells or tissue is preferred, the time has come to consider preparing tissue for the plethora of exciting new molecular and functional assays that have been developed or will be developed in the near future.

One of the most important strategies presented in this chapter employs alternative fixatives for use with cells and tissues. Recent studies described the problems associated with extensive cross-linking of histones to DNA when performing PCR *in situ* hybridization (9). In addition, the effects of aldehyde fixatives on RNA quality and amplifiability have been documented previously (10). These observations were clear in some of our early attempts at *in situ* amplification, so we developed systems to avoid cross-linking fixatives while maintaining acceptable cell and tissue morphology. Using this system, we minimized rigorous digestion to preserve morphology, minimize product diffusion, and maintain antigenicity.

Another recurring theme in this chapter will be the emphasis on fluorescence detection schemes. Using fluorescence for biological studies allows multi-

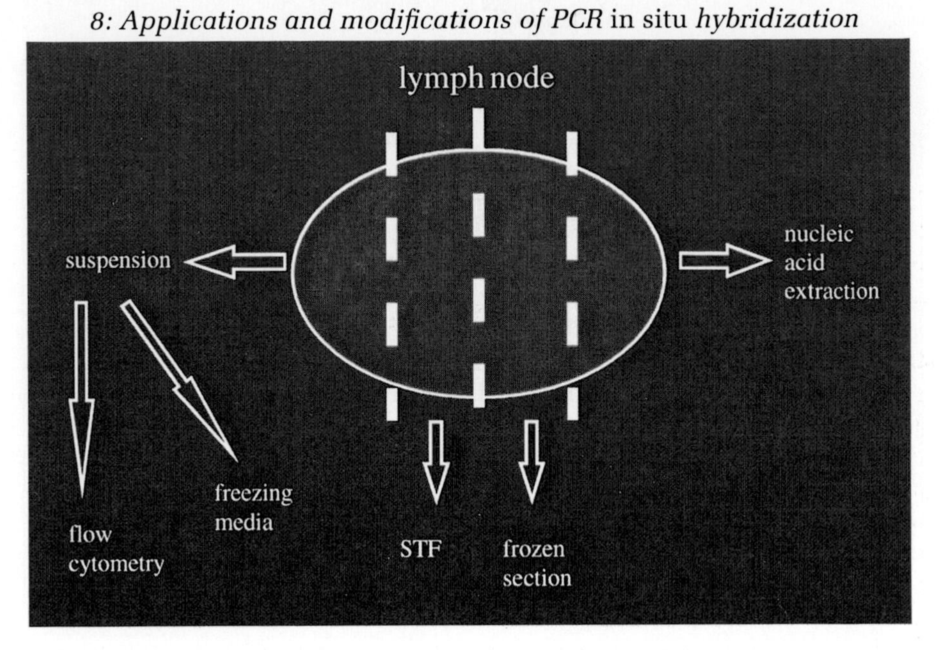

Figure 1. Schematic diagram for the preparation of lymphoid tissue for molecular and immunological analysis.

parameter analysis as one is limited only by the number of colours in the spectrum and our ability to discriminate between them.

Lymphoid tissue has become the tissue of choice for monitoring the effects of antiretroviral therapy on HIV (11). *Figure 1* illustrates a scheme used in our laboratory for preparing lymph nodes which optimizes the tissue for all downstream applications. *Protocol 1* describes our method for preparing cells and tissue for subsequent analyses.

Protocol 1. Preparation of lymphoid cells and tissue for molecular and immunological applications

Equipment and reagents

- Streck Tissue Fixative (STF), molecular biology grade (Streck Laboratories, Inc.)[a]
- Permeafix (Ortho Diagnostics, Inc.)
- Scalpel blades (Fisher)
- Nylon mesh, 37 μm (Fisher)
- Pestle (Fisher)
- RPMI (Gibco)
- 50 ml conical tubes (Falcon, Becton–Dickinson)
- Phosphate-buffered saline (PBS): 130 mM NaCl (Sigma), 10 mM sodium phosphate (Sigma) pH 7.4; store at room temperature
- Tissue embedding compound (Fisher)

Method

1. Divide the lymph node into quarters using a sterile razor blade or scalpel blade.

Protocol 1. *Continued*

2. Wrap one quarter in aluminium foil and snap freeze in isopentane or liquid nitrogen for subsequent nucleic acid extraction.
3. Place another section in aluminum foil, cover in embedding compound, and snap freeze for subsequent *in situ* hybridization or RT *in situ* PCR.
4. Fix the third section overnight (12–24 h) in molecular biology grade STF for subsequent histology, immunohistochemistry, and PCR *in situ* hybridization.
5. Pass the last section through a 37 μm mesh with a pestle and sterile RPMI. Wash the cells twice in PBS: these can then be used immediately for flow cytometric analysis or resuspended in freezing medium for storage.[b]
6. Cells for immunophenotyping cell surface antigens or intracellular antigens can be fixed in Permeafix for 1–72 h with excellent preservation of antigens and nucleic acids including mRNA.

[a]STF penetrates tissue at 2 mm/h whereas 10% neutral buffered formalin penetrates at 2–4 mm/h.
[b]These cells can also be used for T-cell functional assays, for example.

2. Single parameter analysis

2.1 Cells

An abundance of the study of various disease states involves the use of cells and cell lines as *in vitro* models. In addition, blood and peripheral blood mononuclear cells are obtained from study subjects with a minimum of invasiveness. Because of this, we developed all of our *in situ* strategies and protocols on cells first and adapted them for use on tissue as will be discussed later in this chapter. Successful *in situ* techniques on cells are governed by two levels of sensitivity and specificity. Without considering these parameters, the detection of rare cells containing low copy numbers of the target sequence would be difficult. Sensitivity and specificity must be optimized at the single-cell level and in cells within a heterogeneous population. The goal is to detect unambiguously a single copy of a particular target in a single cell. In addition, many applications require the detection of cells containing a single copy of a particular target in a population of cells >90% of which may not contain the particular target (see *Figure 2*). To achieve this goal, we developed a technique (see *Protocol 2*) to amplify a single-copy DNA or low-abundance mRNA (see *Protocol 3*) target to a level where the signal produced allowed unequivocal separation of cells into target-positive and target-negative populations. Our first iteration used the incorporation of digoxigenin-dUTP as a

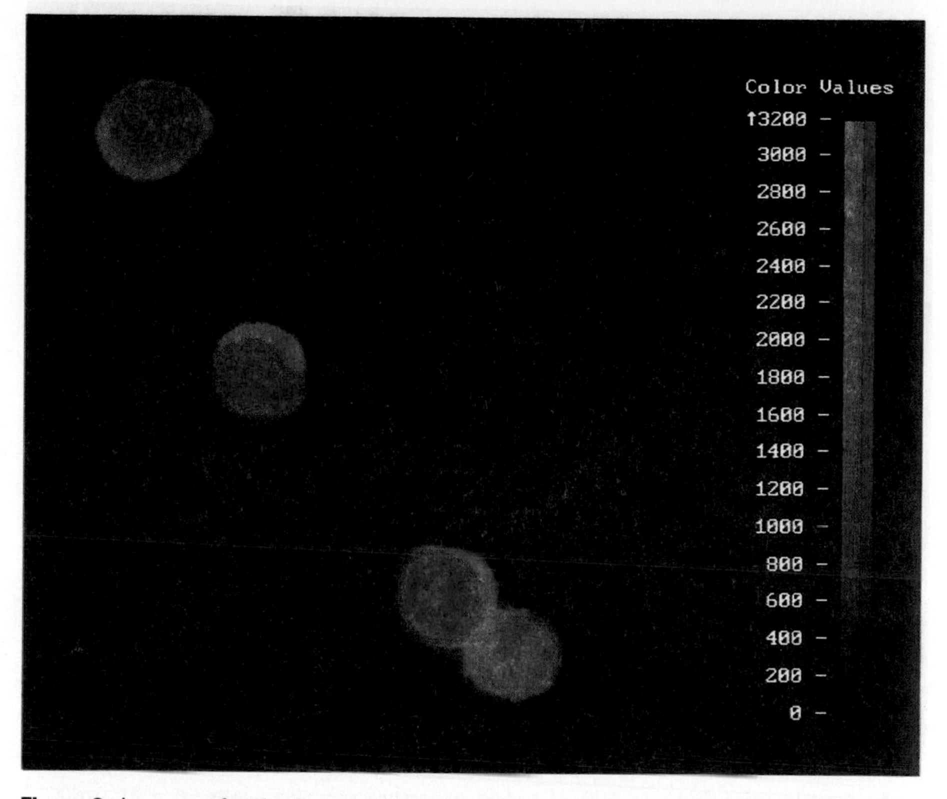

Figure 2. Laser confocal microscope image of HIV-1-infected (yellow–orange) and uninfected (red) cells from an HIV-1-infected individual. HIV-1 proviral DNA was amplified *in situ* and detected using an internally conserved, 5′- and 3′-6-carboxyfluorescein (FAM)-labelled oligonucleotide probe.

reporter. This 'direct incorporation' method yields an unacceptable rate of false positives especially for our goal of detecting rare cells containing our target of interest. Even with the 'hot start' modification we still reserve direct incorporation only for testing digestion conditions, enzyme activity, and cell morphology following thermal cycling.

The detection of low-abundance mRNA deserves special mention. Because of inherent target amplification, one must decide between *in situ* hybridization and RT *in situ* PCR. Issues to be considered are extensive and include:

- sensitivity
- specificity
- quantification

Protocol 2. Detection of single-copy HIV-1 DNA by PCR *in situ* hybridization

Equipment and reagents

- Permeafix (Ortho Diagnostics, Inc.)
- DNA Core PCR Kit (Perkin–Elmer)
- 1 × PCR buffer (10 mM Tris–HCl, pH 8.3; 50 mM KCl)
- PCR reaction mixture (1 × PCR buffer; 2.5 mM $MgCl_2$; 0.25 mM each dATP, dCTP, dGTP; 0.14 mM dTTP; 4.3 μM digoxigenin-11-dUTP; 500 μM each forward and reverse primer; 1.0 μl (10 U) *Taq* DNA polymerase
- Digoxigenin-11-dUTP (Boehringer Mannheim)
- Target specific primers for HIV-1 (437 bp) G51 (5′-CAAATGGTACATCAGGCCATATCACCT-3′) and SK39 (5′-TTTGGTCCTTGTCTTAT-GTCCAGAATGC-3′)
- HIV-1 SK19 probe (5′-ATCCTGGGATTAAATA-AAATAGTAAGAATGTATAGCCCTAC-3′) 5′- and 3′-end-labelled with FAM (Perkin–Elmer). The 5′-end is labelled during synthesis with a FAM phosphoramidite and the 3′-end is labelled with an aminolink and FAM NHS ester (Research Genetics)
- Formamide (Life Technologies, Inc.)
- Salmon sperm DNA, sonicated (Life Technologies, Inc.)
- 2 × SSC/50% formamide/500 μg/ml bovine serum albumin (BSA)
- 1 × SSC/50% formamide/500 μg/ml BSA
- 1 × SSC/500 μg/ml BSA

Methods

1. Centrifuge the cells at 300–600*g* for 2 min and wash the cell pellet twice in PBS.
2. Fix and permeabilize the cells by resuspending with light vortexing in 50 μl of Permeafix and incubate at ambient temperature for 60 min.
3. Centrifuge the cells as above, wash twice with 1 ml PBS, and resuspend the cells in 190 μl of PCR reaction mixture.
4. Amplify the DNA in 500 μl tubes inserted into the wells of a 48-well thermocycler programmed for 25 cycles of thermal denaturation (94°C, 1 min), primer annealing (58°C, 2 min), and primer extension (74°C, 1.5 min), with 5 sec added for each of 25 cycles. Run appropriate positive and negative target controls amplified with or without the addition of *Taq* DNA polymerase simultaneously with each sample.
5. After *in vitro* amplification, centrifuge the cells as above and resuspend in 25 μl of 1 × PCR buffer.
6. Add 100 ng of the appropriately labelled target specific oligonucleotide probe in 10 μg/ml sonicated herring sperm DNA to the reaction tube.
7. Denature the product DNA at 95°C for 3 min then allow the target DNA to hybridize with the respective oligonucleotide probe at 56°C for 2 h.
8. After hybridization, wash the cells with 1 ml of 2 × SSC/50% formamide/500 μg/ml BSA at 42°C, with 1 ml of 1 × SSC/50% formamide/500 μg/ml BSA 30 min at 42°C, and 1 ml of 1 × SSC/500 μg/ml BSA[a] for 30 min at ambient temperature.

[a]Staining of biotinylated antibodies with reporters conjugated to streptavidin can be performed during this step for multiparameter analysis.

In terms of sensitivity, our laboratory uses RT *in situ* PCR only when we have exhausted the most sensitive *in situ* hybridization techniques to detect our target of interest. The reason for this strategy is quantification. To date, *in situ* PCR is not quantitative so estimates of gene expression, which are crucial in most oncological applications, are impossible. Another critical factor in our laboratory is specificity. In HIV pathogenesis, the detection of the HIV regulatory genes *tat*, *rev*, and *nef* is extremely important because of the role of these genes in altering cellular function and in facilitating virion production (12,13). These genes have significant sequence homology and differ only in the utilization of specific splice donors and splice acceptors (14). To detect these genes by routine *in situ* hybridization requires exquisite control of hybridization temperatures and wash stringencies (15). We exploit the differential splicing by designing our probes across splice junctions so that amplification of genomic DNA or, in the case of HIV, genomic RNA will not take place because of the distance between the primer binding sites and the amplification conditions used (4). Following splicing, the primer binding sites are close enough to permit efficient amplification. We follow amplification with *in situ* hybridization using oligonucleotide probes that span the specific splice donor–acceptor sites. Technically, specific downstream primers are preferable to random hexamers for the initial reverse transcription especially when using rTth polymerase as the reverse transcriptase (see *Figure 3*).

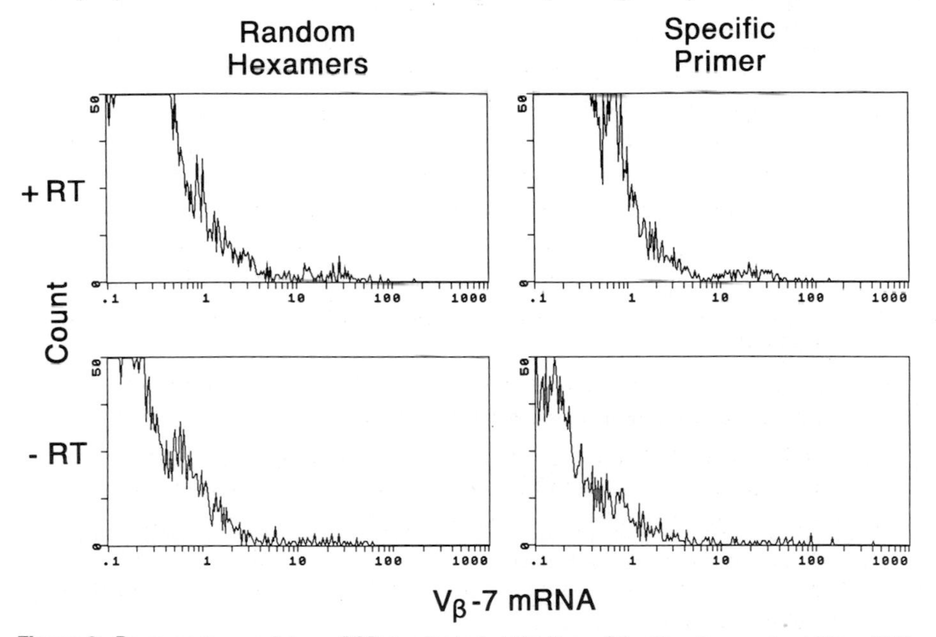

Figure 3. Reverse transcriptase PCR *in situ* hybridization of the T-cell receptor Vβ7 mRNA comparing the difference in signal (+RT) to noise (–RT) using random hexamers or specific downstream primers. Using specific downstream primers results in higher signal-to-noise ratios.

Protocol 3. Detection of multiply spliced HIV-1 *tat* mRNA by RT-PCR *in situ* hybridization

Equipment and reagents

- Permeafix (Ortho Diagnostics, Inc.)
- RNA Core PCR Kit (Perkin–Elmer)
- Reverse transcription reaction mix (1 × RT buffer, 2.5 mM $MgCl_2$, 200 μM each dGTP, dATP, dCTP, 125 μM dTTP, 4 μM digoxigenin-11-dUTP, 10.0 U thermostable rTth polymerase, 40 units RNasin, 500 μM downstream primer)
- Amplification mix (1 × chelating buffer;[a] 200 μM each dGTP, dATP, and dCTP; 125 μM dTTP; 4 μM dUTP-11-digoxigenin; 500 μM upstream primer; 5 units *Taq* polymerase[b])
- Digoxigenin-11-dUTP (Boehringer Mannheim)
- Primers (upstream, 5′-GCGAATTCATGGAG/TCCAGTAGATCCTAGACTA-3′; downstream, 5′-GCTCTAGACTATCTGTCCCCTCAGCTACTGC-TATGG-3′)
- *tat* splice junction probe (5′-TTCTCTAT-CAAAGCAACCCACCTCCCAATC-3′) labelled at the 5′- and 3′-ends with FAM (Perkin–Elmer)
- 2 × SSC/50% formamide/500 μg/ml BSA
- 1 × SSC/50% formamide/500 μg/ml BSA
- 1 × SSC/500 μg/ml BSA

Methods

1. Following Permeafix treatment and washes (see *Protocol 2*), resuspend the cells in 40 μl of reverse transcription reaction mixture.
2. Incubate samples for 10 min at 70 °C and then place on ice.
3. Add 160 μl of PCR reaction mixture directly to the reverse transcription mixture.
4. Amplify the samples as described above (see *Protocol 2*).
5. After *in vitro* amplification, centrifuge the cells as above and resuspend in 25 μl of 1 × PCR buffer.
6. Add 100 ng of the appropriately labelled target specific oligonucleotide probe in 10 μg/ml of sonicated herring sperm DNA to the reaction tube.
7. Denature the product DNA at 95 °C for 3 min then allow the target DNA to hybridize with the respective oligonucleotide probe at 56 °C for 2 h.
8. After hybridization, wash the cells with 1 ml of 2 × SSC/50% formamide/500 μg/ml BSA at 42 °C, with 1 ml of 1 × SSC/50% formamide/500 μg/ml BSA for 30 min at 42 °C, and 1 ml of 1 × SSC/500 μg/ml BSA[a] for 30 min at ambient temperature.

[a] Contains EGTA which selectively chelates manganese. This two-step reaction allows the control of the magnesium concentration in the amplification step which is a critical parameter for successful PCR *in situ* hybridization.
[b] May not be necessary with all applications. PCR *in situ* hybridization, in general, requires twice as much polymerase as solution-phase PCR.

2.2 Tissue

Many protocols have already been presented to perform PCR *in situ* hybridization for a single target. PCR *in situ* hybridization is extremely application-specific with target-to-target and even tissue block-to-block variation. These phenomena make it difficult to learn this technique and to build confidence in performance. Before proceeding to the more sophisticated, multiparameter variations of PCR *in situ* hybridization, a useful control tissue system will be described (see *Protocol 4*). The gene for HLA-DQα is found in all nucleated cells in humans (see *Figure 4*). Using a commercially available

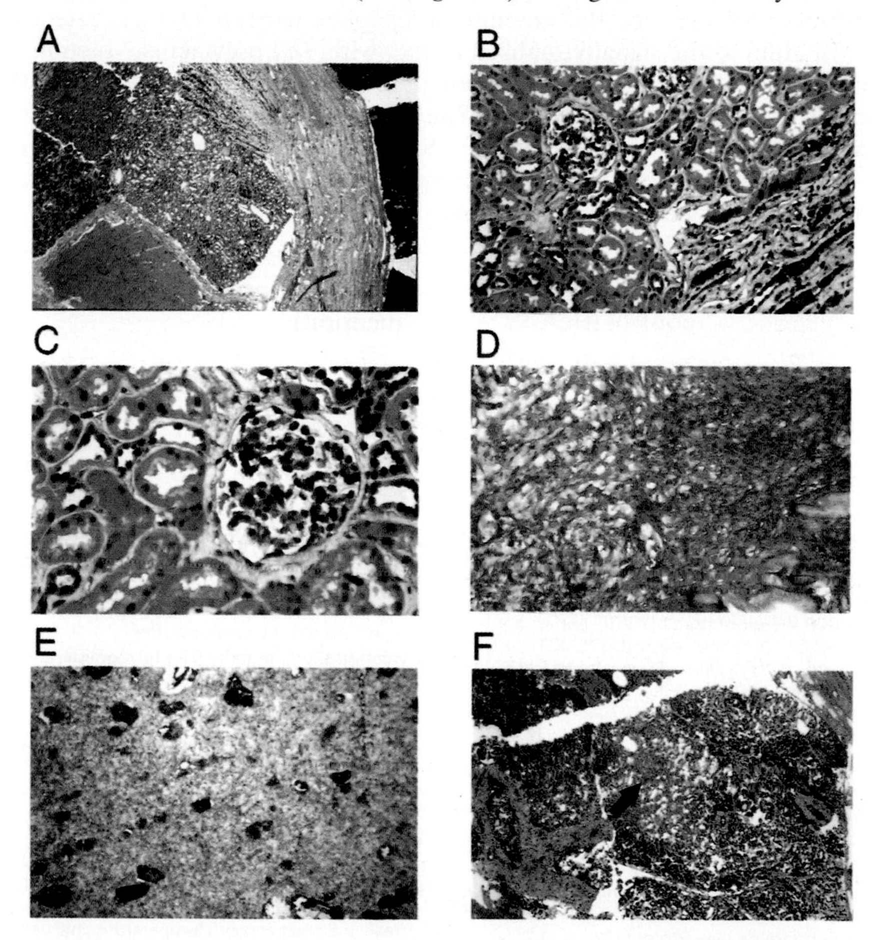

Figure 4. PCR *in situ* hybridization for HLA-DQα in a normal tissue sausage (BioGenex) consisting of 15 different human tissues. Amplicon DNA was probed with an internally conserved, digoxigenin-tailed oligonucleotide probe. The hybrids were detected with anti-digoxigenin antibody conjugated with alkaline phosphatase and the substrate NBT/BCIP. Representative light micrographs of (A) tissue sausage after PCR *in situ* hybridization for HLA-DQα, ×40; (B) renal tubules and glomeruli, ×100; (C) renal tubules and glomeruli, ×200; (D) normal tissue sausage amplified with nonsense primers and probed with an HLA-DQα oligonucleotide probe, ×100. Note the diffuse purple colour due to amplicon diffusion secondary to over-digestion in (E) and the diminished counterstain (arrow) of a bubble in (F).

multi-tissue sausage (BioGenex), one can use the following protocol for practice or optimization of protease digestion on tissue that may represent a specific tissue of interest. By adhering three sausage sections per slide, one can perform both *in situ* PCR and PCR *in situ* hybridization with the appropriate controls. For *in situ* PCR, sections with *Taq* polymerase, without *Taq* polymerase, and without primer are most appropriate. The 'no-primer' section allows the researcher to visualize the amount of template-dependent nick extension contributing to the signal visualized in the 'with *Taq* polymerase' section. In general the number of cells undergoing nick extension is much lower than the number of positive cells in the 'with *Taq* polymerase' section, verifying that at least some signal is true amplification. Since nicks can be induced by fixation or as a consequence of natural processes such as apoptosis, this 'no-primer' control can provide useful information.

Protocol 4. PCR *in situ* hybridization for a representative tissue control (HLA-DQα amplification)

Equipment and reagents

- GeneAmp *In Situ* PCR System 1000 containing a thermal cycler, Assembly Tool, Disassembly Tool, AmpliCover Disc and Clips and glass slides (silane-coated, 1.2 mm thick; Perkin–Elmer, see Chapter 9).
- GeneAmp *In Situ* PCR Core Kit containing AmpliTaq DNA Polymerase IS, 10 × PCR Buffer, $MgCl_2$, and 10 mM dNTPs (Perkin–Elmer)
- PCR reaction mix (1 × PCR buffer; 3.5 mM $MgCl_2$; 0.25 mM each dATP, dCTP, dGTP, and dTTP; 500 μM each forward and reverse primer).
- Normal Tissue Sausage Slides (BioGenex)
- Coplin jars (Fisher)
- Histology grade 100% ethanol (Fisher)
- Xylene (BDH P/N C4330 or equivalent)
- Proteinase K (Boehringer Mannheim)
- Coverslips (Fisher)
- Digoxigenin-11-dUTP (see *Protocol 2*)
- Streptavidin, alkaline phosphatase-conjugated, 750 U/mL (Boehringer Mannheim)
- Formamide (see *Protocol 2*)
- Salmon sperm DNA, sonicated (see *Protocol 2*)
- Humid chamber (Sigma)
- Hot temperature block (Fisher)
- Phosphate-buffered saline (PBS): 130 mM NaCl, 10 mM sodium phosphate pH 7.4; store at room temperature
- Lysis buffer; 20 mM Tris–HCl (Sigma) pH 7.4, 0.5% SDS (Gibco); store at room temperature
- Proteinase K solution (Boehringer Mannheim): 10 mg/mL in lysis buffer
- Hybridization solution: 50% formamide, 2 × SSC, 200 μg/mL sheared salmon sperm DNA, 50–250 μg biotin-labelled oligonucleotide probe (Research Genetics)
- Buffer 3: 100 mM Tris–HCl pH 9.5, 100 mM NaCl, 50 mM $MgCl_2$
- 4-Nitroblue tetrazolium chloride (NBT) solution (Boehringer Mannheim)
- 5-Bromo-4-chloro-3-indoyl-phosphate (BCIP) solution (Boehringer Mannheim)
- NBT/BCIP substrate solution: add 45 μl NBT and 35 μl BCIP to 10 mL buffer 3; prepare fresh and store at room temperature until use
- Slide mounting medium (Fisher)
- Fast Green stain (Sigma)

Methods

1. Deparaffinize the tissue sections through xylenes for 10 min, 100% ethanol for 10 min, 95% ethanol for 10 min.
2. Air dry[a] and rehydrate in PBS for 5 min.
3. Digest with 20 μg/ml proteinase K in 1 × lysis buffer (20 mM Tris–HCl

pH 7.4, 0.5% SDS) (must be titred) for 1 h at 37°C. Use 500 μl of the digestion mixture per slide. Cover the mixture gently with a coverslip that covers the length of the tissue sections.

4. Inactivate the proteinase K for 1 min at 95°C by placing the slide with the coverslip on the heat block.
5. Wash in PBS for 1 min by placing the slide with coverslip into the buffer.[b]
6. Dehydrate through graded ethanols (80%, 95%, and 100%) for 5 min in each dilution, then air dry as previously described.
7. Apply prewarmed (70°C) reaction mix (50μl) to the tissue section placed on the Assembly Tool.
8. Add 10 units of AmpliTaq IS to the buffer bead[c] on the positive control section. Create the sealed chamber using the Assembly Tool. Place the slide vertically in the GeneAmp 1000.
9. Amplify for 30 cycles (94°C for 1 min, 56°C for 2 min, 72°C for 2 min with a 15°C soak).
10. Disassemble the slides with the Disassembly Tool.
11. Wash once in PBS for 1 min.
12. Dehydrate through graded ethanols (50%, 80%, 95%, and 100%) for 5 min each.
13. Apply 50 μl of hybridization mix (5 × SSC, 50% formamide, 0.5% Tween 20, 100 μg/ml sonicated salmon sperm DNA) with 75 ng/section[d] digoxigenin- or biotin-tailed probe. Cover the slides with a coverslip.
14. Denature the slides for 2 min at 92°C.
15. Incubate the GeneAmp 1000 overnight at 37°C.
16. Wash the slides in 2 × SSC, 0.5% Tween 20 for 10 min, 0.2 × SSC, 0.5% Tween 20 for 10 min at room temperature.
17. Add 500 μl of a 1/100 dilution of anti-digoxigenin antibody to the slides, cover each slide with a coverslip, and incubate at 4°C for 4 h.
18. Wash twice in PBS for 5 min.
19. Incubate in buffer 3 for 2 min.
20. Detect with NBT/BCIP diluted in buffer 3 in the dark. Monitor the development of the purple precipitate using light microscopy.
21. Counter-stain with Fast Green for 30 sec or Nuclear Fast Red for 3–5 min and mount in Crystal Mount or other aqueous mountant.

[a] Tissue appears pure white rather than grey when the tissue is appropriately air dried.
[b] Allow the coverslip to fall off the slide passively. If the coverslip does not fall off within a few seconds, gently dip the slide up and down in the PBS.
[c] If the buffer bead is off-centre, place the polymerase aliquot over the centre of the tissue section and the buffer bead will migrate back towards centre. This technique will minimize bubble formation.
[d] The probe must be titred; use the concentration of probe that does not give signal when hybridized to an uncycled tissue of interest.

3. Multiparameter analysis

The analysis of multiple phenotypic and/or genotypic markers in a particular cell or tissue is extremely important from a biological and a technical standpoint. Biologically, altered protein expression can be found in a variety of disease states and certainly in HIV infection where viral gene expression can have profound effects on the expression of various cell surface molecules (12,13). Technically, the simultaneous use of a phenotypic marker with *in situ* hybridization or PCR *in situ* hybridization helps verify specificity when target genes are found in specific sub-populations of cells (see *Figure 5*). Similarly, if the target gene is found in a rare sub-population of cells, simultaneous immunophenotyping with another fluorescent dye can enrich for the sub-population of interest without a preparative physical separation.

Simultaneous immunophenotyping and *in situ* hybridization have been used extensively for a number of years. Simultaneous immunophenotyping and PCR *in situ* hybridization have recently been performed (13,16,17). Two strategies have been employed. In the first iteration (16,17), routine immunophenotyping was performed on the tissue prior to PCR *in situ* hybridization. Consideration of an enzymatic substrate that would survive subsequent

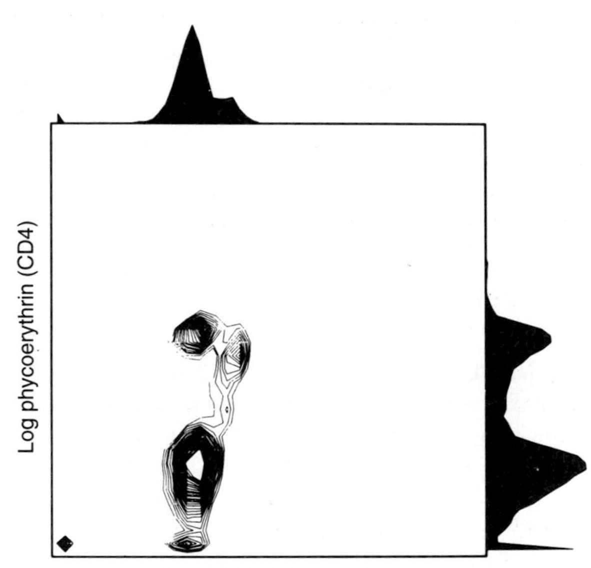

Figure 5. Dual immunophenotyping and PCR-driven *in situ* hybridization was performed on monocyte-depleted peripheral blood mononuclear cells using the anti-CD4 antibody, Leu3A. Three populations are shown in this representative topogram: the $CD4^-$, HIV^- population (lower left), the $CD4^+$, HIV^- population (upper left), and the $CD4^+$, HIV^+ population (upper right).

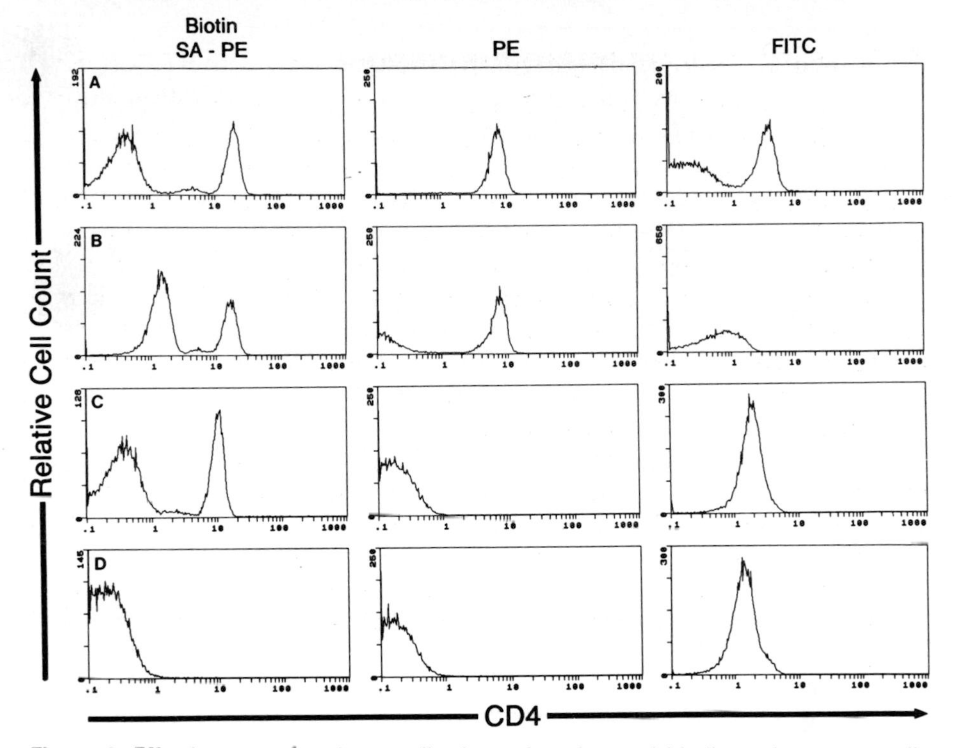

Figure 6. Effectiveness of various antibody conjugation and binding schemes on cells subjected to thermal amplification *in situ*. The histograms in row A were generated using cells with antibody bound prior to fixation and permeabilization. The histograms in row B were generated using cells with antibody bound following fixation and permeabilization but before thermal amplification. The histograms in row C were generated using cells with antibody bound before fixation and permeabilization followed by thermal amplification. In row C, streptavidin–phycoerythrin was added after thermal amplification and hybridization. The histograms in row D were generated using cells with antibody bound following fixation, permeabilization, and thermal amplification. SA, streptavidin; PE, phycoerythrin; FITC, fluorescein isothiocyanate.

alcohol dehydrations and rehydrations was given and diaminobenzidine (DAB) was shown to be best for this particular strategy. Since we were interested in fluorescence detection with flow or image analysis, our laboratory investigated the resilience of antibodies and antibody conjugates to thermal cycling (13). *Figure 6* illustrates experiments which verified the thermal stability of biotinylated antibodies to thermal cycling. In addition, these data also showed that antibodies bound to antigen prior to thermal cycling had a protective effect on the antigen as immunophenotyping following thermal cycling was unsuccessful with any antibody conjugation. *Protocol 5* describes the modifications of *Protocols 2* and *3* for dual immunophenotyping PCR and RT-PCR *in situ* hybridization.

Protocol 5. Immunophenotyping simultaneously with RT-PCR *in situ* hybridization, DNA-PCR *in situ* hybridization, and *in situ* hybridization

Equipment and reagents

- Antibodies (Becton-Dickinson)
- PBS, pH 7.4 and 8.3 (see *Protocol 2*)
- Permeafix (see *Protocol 2*)
- Streptavidin-conjugated fluors (Becton-Dickinson)

Methods

1. Adjust cell samples to a concentration of 1–2 $\times$ 10^6 cells per ml in PBS, pH 7.4.
2. Centrifuge the cells at 300–600*g* for 5 min.
3. Remove the supernatant and resuspend the cells in 90 μl of PBS and 10 μl of biotinylated anti-CD4 or CD14, for example.[a]
4. Centrifuge the cells at 300–600*g* for 2 min and wash the cell pellet twice in PBS.
5. Fix and permeabilize the cells by resuspending with light vortexing in 50 μl of Permeafix and incubate at ambient temperature for 60 min.
6. Centrifuge the cells as above, wash twice in 500 μl PBS, pH 7.4, resuspend in PCR reaction mix and continue with *Protocol 2, 3,* or *6*.
7. Following the last wash in *Protocols 2* and *3* or following the post-FISNA PBS wash (*Protocol 6*, Step 7), resuspend the cells in 80 μl of PBS, pH 7.4, and 20 μl streptavidin–phycoerythrin and incubate the mixture for 30 min at ambient temperature. Wash the cells in PBS, pH 7.4, as described above.
8. Centrifuge the cells as described and resuspend in PBS, pH 8.3, for analysis by flow cytometry or image analysis.[b]

[a] Dinitrophenol (DNP) conjugated antibodies are also thermostable.
[b] Cells can be attached to slides for image analysis using a cytocentrifuge. A total of 2.5 $\times$ 10^4–1.0 $\times$ 10^5 cells is optimal for each spot on the slide.

4. Fluorescence *in situ* 5′-nuclease assay and fluorescence detection

The detection of antigens and genes using fluorescently labelled antibodies and genes has become a powerful tool in diagnostics and disease pathogenesis research. The advantages of fluorescence applications include eliminating radioactivity from the laboratory, ease of use, and, most importantly, multiparameter capabilities. The disadvantages of fluorescence include relative insensitivity when compared with radioactivity, expense, and autofluorescence in cells and tissues. In designing fluorescent probes for *in situ* hybridization, PCR *in situ* hybridization, and fluorescence *in situ* 5′-nuclease assay (FISNA), many factors must be considered. Although probe design will depend somewhat on the particular application, we find that oligonucleotides are the most versatile for *in situ* applications. The advantages of oligonucleotides include size, ease of commercial or custom synthesis, purity, and labelling options. The disadvantages of oligonucleotides are relatively easy to overcome. Because oligonucleotides are short, the number of labels per probe is limited. This shortcoming can be conquered by using cocktails of oligonucleotides labelled on the 5′- and 3′-ends. We used over two hundred 25mers to detect low-abundance HIV mRNA (18) in one study. Cocktails of oligonucleotides also minimize the effects of non-specific hybridization or cross-hybridization by a particular oligonucleotide in the cocktail. Longer oligonucleotides (>30 bases) also decrease the chances of cross-hybridization with genomic sequences.

The choice of fluorescent labels on an oligonucleotide depends on the application, the use of other fluors on ligands or antibodies, and the availability of a particular fluor in a chemical configuration consistent with the synthesis. Fluors are available as deoxyribonucleotides or ribonucleotides (e.g. dUTP, dCTP), dideoxynucleotides, NHS esters, and phosphoramidites, for example. Oligonucleotides can be labelled during synthesis (e.g. phosphoramidites, deoxyribonucleotides, or ribonucleotides) or after synthesis (e.g. NHS esters). The hallmark of a successful probe is purity. Oligonucleotides synthesized on an automated synthesizer must be purified by HPLC, for example, to minimize unlabelled probe and free aminolinks which can bind non-specifically to cells.

The most important determinant of a successful experiment involving fluorescence is signal-to-noise ratio. This ratio will govern the resolution of cells containing a particular gene from cells lacking that gene. In other words, the signal-to-noise ratio determines the sensitivity of a particular assay. The signal can be increased by increasing the number of labels per probe, by increasing the number of probes containing a set number of labels, or by maximizing the intensity of dye fluorescence. The first two choices are straightforward; maximizing the intensity of dyes can be achieved by adjusting the excitation wavelength to approximate the excitation maximum of the particular dye. Just because a flow cytometer has an argon laser, the excitation wavelength does not have to be 488 nm. Newer instruments have lasers that can be easily adjusted to provide excitation at wavelengths closer to the excitation maxima of various dyes. *Table 1* shows the excitation and emission maxima of dyes used for probe synthesis in our laboratory.

Noise in fluorescence analysis of cells and tissue has been one of the major impediments in the widespread use of this technology. Specifically, autofluorescence in the fluorescein emission range is most disconcerting in many of the

Table 1 Representative fluorescent dyes used for probe synthesis

Dye	Chemistry	Excitation (nm)	Emission (nm)	Optimal laser[a]
Fluorescein	dNTP, ddNTP, UTP, amidite	494	517	argon (488)
FAM[b]	amidite	494	518	argon (488)
TET[c]	amidite	521	538	argon (514)
HEX[d]	amidite	535	553	argon (528), krypton
TAMRA[e]	NHS[f] ester, dNTP	560	582	krypton
ROX[g]	NHS ester	587	607	krypton
Texas Red	ddNTP	593	612	krypton
Cy5	amidite	646	672	krypton

[a] Laser lines are indicated in parentheses
[b] FAM: 6-carboxyfluorescein
[c] TET: tetrachloro-6-carboxyfluorescein
[d] HEX: hexachloro-6-carboxyfluorescein
[e] TAMRA: tetramethyl-6-carboxyrhodamine
[f] NHS: N-hydroxysuccinimide
[g] ROX: 6-carboxy-X-rhodamine

common applications. This problem can be avoided by using dyes that emit outside the range of maximal autofluorescence or by using autofluorescence quenching dyes such as Evan's Blue or Trypan Blue. In addition, the thermal cycling of cells and tissue increases the background or autofluorescence by five times the number of fluorescein equivalents (19).

Most importantly, the use of fluorescence allows interfacing with other instruments to provide additional or confirmatory data. For example, we verified the intracellular amplification and detection of HIV-1 DNA (5) by extensively washing cells to remove amplified product in the supernatant then lysing the cells and precipitating the fluorescence heteroduplex created *in situ*. This DNA was analysed on a laser sequencer (Perkin–Elmer 373, Foster City, CA) fitted with GeneScan 672 software (see *Figure 7*). This analysis revealed a single peak of the appropriate size which was calculated using internal molecular weight markers labelled with another compatible fluorescent dye (ROX). In addition, we used image analysis to verify flow cytometry results and vice versa. Using dual immunophenotyping and *in situ* hybridization, we were able to detect HIV-1 RNA in CD83-positive dendritic cells in peripheral blood following adherence depletion and CD14 panning. Because conflicting reports debate the presence of HIV in dendritic cells versus in T-lymphocytes adhering to dendritic cells, we performed Z-banding analysis on CD83-positive, HIV-1-RNA-positive cells sorted by flow cytometry (see *Figure 8*).

Taq polymerase, the enzyme most commonly used for gene amplification, has 5′-nuclease activity capable of releasing mononucleotides or dinucleotides into solution during extension through double-stranded regions (20). In addition, rTth polymerase also has 5′-nuclease activity allowing amplification of RNA targets using single buffer systems (B.K. Patterson, unpublished data).

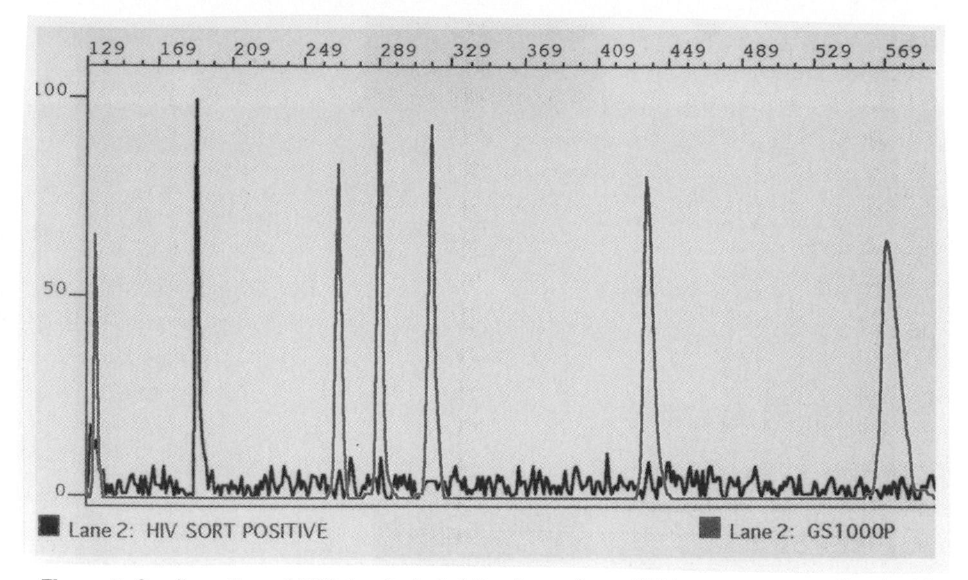

Figure 7. Confirmation of PCR *in situ* hybridization using a 373 laser sequencer fitted with GeneScan 672 software. Cells from the HIV-1-positive and -negative populations were sorted, washed, and lysed by incubation at 56°C for 45 min with 100 µg/ml proteinase K, and run on a 6% non-denaturing polyacrylamide gel. Note the peak in the HIV-positive sort-positive trace.

Applications have been developed using this enzymatic activity to detect amplified product (21) in solution. Instead of detecting signal using a fluorescence spectrometer as in previous studies, FISNA signal can be detected within cells using a standard monochromatic fluorescence microscope, a laser confocal microscope, or a flow cytometer (see *Figure 9*) (22).

Probe design is critical for successful FISNA. Although a 5′ quenching dye and a 3′ reporter dye (TFSK19) might be more desirable for *in situ* applications because of the bulkier group bound to the reporter, the fluorescence characteristics of the probe may be suboptimal (see *Figure 10*) (22). Effective quenching of the reporter dye is dependent on the approximation, within the critical Forster distance (23), of the reporter dye and the quencher dye. The 5′ TAMRA/3′ FAM (TF) orientation of reporter and quencher did not allow an appropriate configuration of the HIV-1 oligonucleotide probe SK19 in solution for effective quenching of the reporter.

Protocol 6 describes the use of STF, a non-cross-linking fixative, and Permeafix, a detergent-based permeabilizing agent eliminating protease digestion. Protease digestion requires re-optimization for each new application and even each new block. Over-digestion can lead to product diffusion and poor morphology, for example. Under-digestion can diminish signal intensity. STF minimizes cross-linking of nucleic acids and proteins and Permeafix maintains antigenicity without over-digesting. Previous studies have shown that HIV mRNA can be

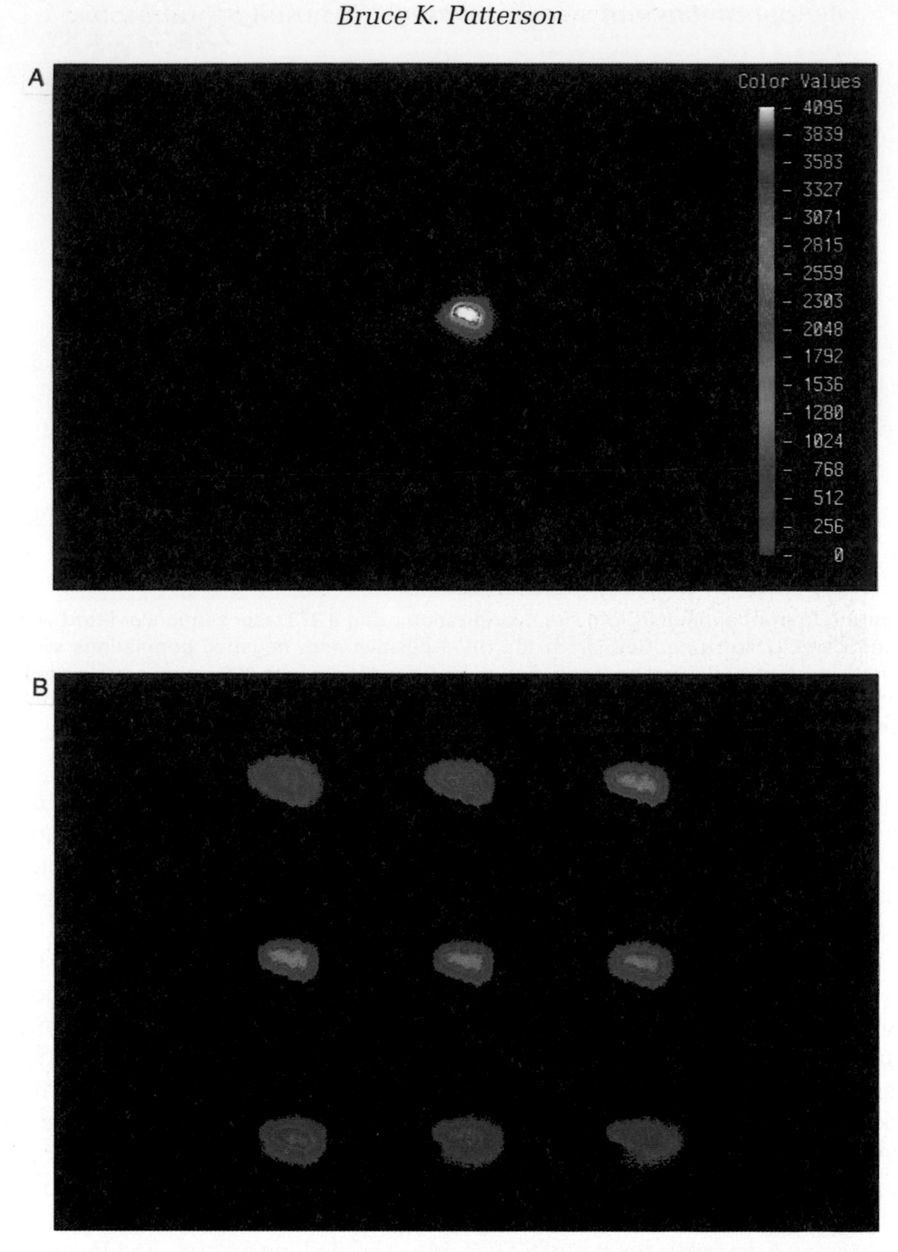

Figure 8. Verification of HIV gene expression in peripheral blood dendritic cells by fluorescence-activated cell sorting (A) and Z-banding analysis by laser confocal microscopy (B). Z-banding confirmed the intracellular localization of HIV RNA.

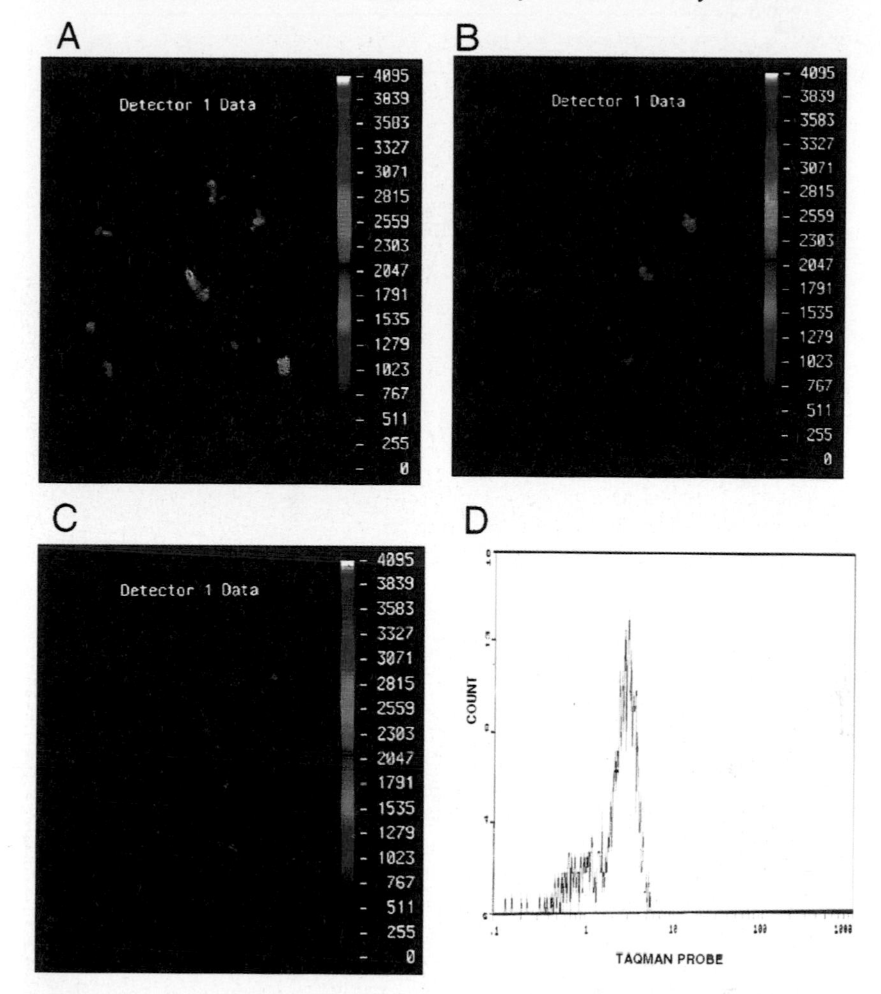

Figure 9. Laser confocal images of HIV-1-infected and uninfected cells following FISNA analysis for HIV-1 DNA. (A) HIV-1 DNA detected in CEM/IIIB cells using FISNA and AmpliTaq IS polymerase; (B) lack of detection of HIV-1 DNA in CEM/IIIB cells using FISNA and *Taq* CS polymerase, which lacks 5′ nuclease activity; (C) lack of detection of HIV-1 DNA in CEM cells using FISNA and AmpliTaq IS polymerase; (D) detection of cells containing HIV-1 DNA in a 3:1 mix of CEM/IIIB:CEM by FISNA and flow cytometry.

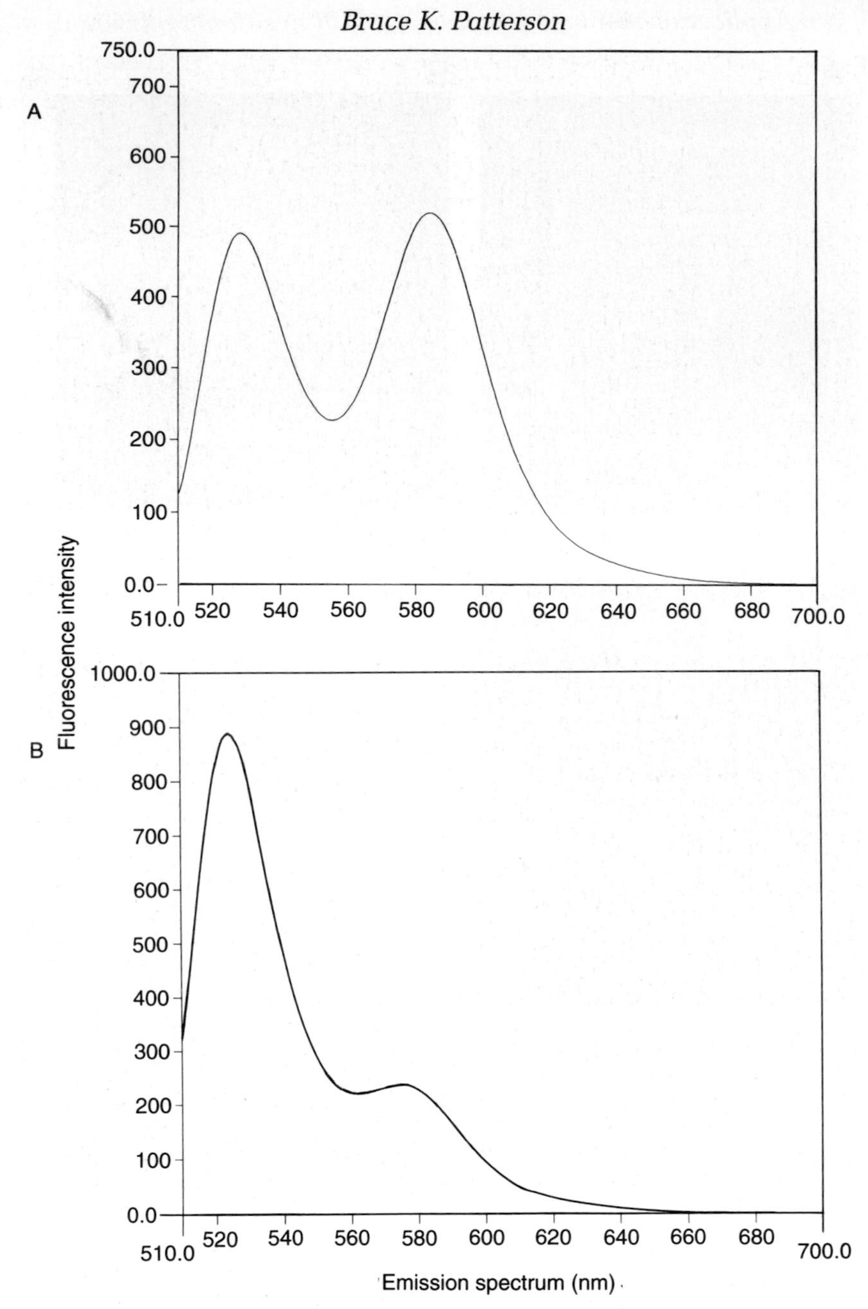

Figure 10. Spectrofluorimetric analysis of SK19 FT (A) and SK19 TF (B) probe emission spectra at 488 nm excitation in Tris–EDTA buffer. The emission peak at 518 nm reflects FAM emission (reporter) and the peak at 582 nm reflects 5-methyl-6-carboxyrhodamine emission (quencher). The ratio of reporter intensity to quencher intensity (*R*/*Q*) reflects the efficiency of quenching in solution. Lack of FAM quenching in the SK19 TF probe limits the utility of this probe in FISNA.

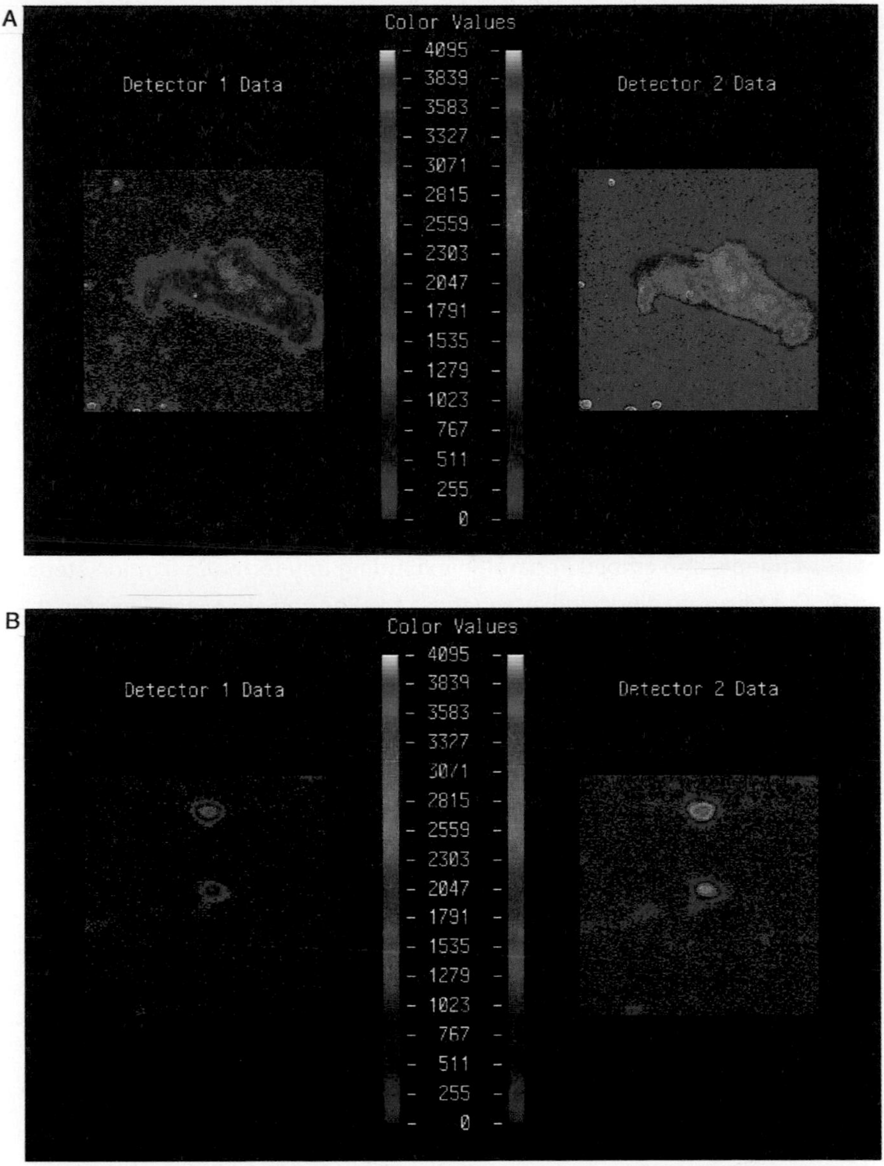

Figure 11. Detection of CD4-positive HIV-1 DNA positive cells by dual immunophenotyping/FISNA on a cervicovaginal lavage specimen from an HIV-seropositive patient. (A) Low magnification image of epithelial and inflammatory cells analysed for CD4 expression (detector 2) and HIV-1 DNA (detector 1). (B) Higher power image of two CD4-positive cells (right), one of which is HIV-1-DNA-positive (left, top cell).

detected in cells fixed with Permeafix for 1-72 h (B.K. Patterson, submitted). Lastly, *Protocol 5* can be combined with *Protocol 6* for dual immunophenotyping/FISNA in cells (see *Figure 11*) with results obtained in a single day (22).

Protocol 6. Detection of HIV-1 DNA using FISNA

Equipment and reagents

- STF (see *Protocol 2*)
- Permeafix (see *Protocol 2*)
- Xylene (see *Protocol 2*)
- Ethanol (see *Protocol 2*)
- Primers (see *Protocol 2*)
- Probe (FAM-5′-CCTGGGATTAAATAAAAT-AGTAAGAATGTATAGCCCTACp-3′-TAMRA)
- Trypan Blue solution (1.0 μg/ml in PBS, pH 7.4)
- 5′-Nuclease assay reaction mix (1 × PCR buffer, 2.5 mM $MgCl_2$, 200 mM dATP, 200 mM dCTP, 200 mM dGTP, 400 mM dUTP, 0.5 U AmpErase uracil *N*-glycosylase, and 10 U AmpliTaq IS or *Taq* CS polymerase) (Perkin–Elmer)

Method

1. Cut the tissue samples to 0.5 cm thickness and fix for at least 18 h in molecular biology grade STF. Process the tissue through graded ethanols and embed in paraffin excluding any formalin fixation steps.
2. Cut tissue sections and attach them to silanized slide (see Chapter 3).[a]
3. Treat the cells and tissue sections with Permeafix for 1 h at room temperature and wash the sections twice in PBS.[b]
4. Dehydrate the cells and tissue sections through graded ethanols and air dry prior to thermal amplification as previously described (see *Protocol 2*).
5. Add 50 μl of 5′ nuclease reaction mixture per cell smear or tissue section. Assemble the slides using the Assembly Tool of the GeneAmp 1000 *in situ* PCR system.
6. Cycle the slides using the following profile: 50 °C for 2 min, 92 °C for 2 min followed by 30 cycles consisting of 92 °C for 1 min, 56 °C for 2 min, and 72 °C for 2 min, followed by a 4 °C soak. Perform no-*Taq*-polymerase and no-primer controls on each slide. Slides with tissue and cells lacking the desired target should be prepared identically and included with each run.
7. Following amplification, wash the sections in PBS for 5 min at ambient temperature.
8. Fix the sections in 4% paraformaldehyde for 5 min at room temperature.
9. Wash the slides twice in PBS, pH 8.3 and counter-stain in Trypan Blue solution.
10. Place coverslips on the sections and analyse the tissue using a laser confocal microscope or monochromatic light microscope.

[a] If tissue tends to fall off the slides, place one drop of 3% (v/v) white glue in water on the slide prior to adding the tissue sections. Without floating the section in a water bath, place the

section on the glue bead and heat for a few seconds at 42°C or until the tissue flattens. Do not deparaffinize. Blot excess glue with gauze. Allow to dry for at least 2 h.
[b]Proteinase K can be substituted when using tissue fixed in cross-linking fixatives.

In summary, the techniques presented in this chapter and other chapters in this book introduce the era of single-cell biology. We can analyse the effects of disease on individual cells with the hope of restoring normal function at the most basal level. I have introduced the concept of fluorescence as a unifying reporter for the multiparameter analysis of cells and tissues. In addition, we use instruments to detect many of the same dyes used *in situ* for sequencing, heteroduplex mapping, differential display, and liquid hybridization applications. Because fluorescent dyes are relatively non-toxic and assays are easily automated, applications employing fluorescent detection should continue to move into the realm of clinical diagnosis.

References

1. Popovic, M., Sarugadharan, M.G., Read, E., and Gallo, R.C. (1984). *Science*, **224**, 497.
2. Pomerantz, R.,Trono, D., Feinberg M.B., and Baltimore, D. (1990). *Cell*, **61**, 1271.
3. Haase, A.T., Retzel, E.F. and Staskus, K.A. (1990). *Proc. Natl Acad. Sci. USA*, **87**, 4971.
4. Patterson, B.K., Till, M., Otto, P., Goolsby, C., Furtado, M.R., McBride, L.J., and Wolinsky, S.M. (1993). *Science*, **260**, 976.
5. Pantaleo, G., Graziosi, C., and Fauci, A.S. (1993). *New Engl. J. Med.*, **328**, 327.
6. Ho, D.D., Neumann, A.U., Perelson, A.S., Chen, W., Leonard, J.M., and Markowitz, M. (1995). *Nature*, **373**, 123.
7. Wei, X., Ghosh, S.K., Taylor, M.E., Johnson, V.A., Emini, E.A., Deutsch, P. *et al.* (1995). *Nature*, **373**, 117.
8. Koffron, A.J., Mueller, K.H., Kaufman, D.B., Stuart, F.P., Patterson, B.K., and Abecassis, M.I. (1995). *Scand. J. Infect. Dis., Suppl.*, **99**, 61.
9. Teo, I.A. and Shaunak, S. (1995). *Histochem. J.*, **27**, 660.
10. Klimecki, W.T., Futscher, B.W., and Dalton, W.S. (1994). *BioTechniques*, **16**, 1021.
11. Haase, A.T., Henry, K., Zupancic, M., Sedgewick, G., Faust, R.A., Melroe, H. *et al.* (1996). *Science*, **274**, 985.
12. Garcia, J.V., Alfano, J., and Miller, A.D. (1993). *J. Virol.*, **67**, 1511.
13. Patterson, B.K., Goolsby, C., Hodara, V., Otto, P., Lohman, K., and Wolinsky, S.M. (1995). *J. Virol.*, **69**, 4316.
14. Furtado, M.R., Baalachandran, R., Gupta, P., and Wolinsky, S.M. (1990). *Virology*, **185**, 258.
15. Peng, H., Reinhart, A., Retzel, E.F., Staskus, K.A., Zupancic, M., and Haase, A.T. (1994). *Virology*, **206**, 16.
16. Embretson, J., Zupancic, M., Ribas, J.L., Burke, A., Racz, P., Tenner-Racz, K., and Hasse, A.T. (1993). *Nature*, **362**, 359.
17. Gressens, P. and Martin, J.R. (1994). *J. Neuropathol. Exp. Neurol.*, **53**, 127.
18. Patterson, B.K. *Cytometry* (in press).
19. Mosiman, V. *Cytometry* (in press).

20. Holland, P.M., Abramson, R.D., Watson, R., and Gelfand, D.H. (1991). *Proc. Natl. Acad. Sci. USA*, **88**, 7276.
21. Bassler, H.A., Flood, S.J.A., Livak, K.J., Marmaro, J., Knorr, R., and Batt, C.A. (1995). *Appl. Environ. Microbiol.*, **61**, 3724.
22. Patterson, B.K., Jiyampa, D., Mayrand, P.E., Hoff, B., Abramson, R., Garcia, P.M., *et al.* (1996). *Cancer Res.*, **24**, 3656.
23. Mergny, J.-L., Boutorine, A.S., Garestier, T., Belloc, F., Rougee, M., Bulychev, N.V., *et al.* (1994). *Nucleic Acids Res.*, **22**, 920.

9

Automation of *in situ* amplification

1. Technologies available for *in situ* amplification

JOHN J. O'LEARY

In situ polymerase chain reaction (PCR) amplification has been carried out using standard thermocyclers, heating blocks, or cycling ovens, none of which was specifically designed to perform these techniques. Initially most investigators covered a standard multi-well PCR block with an aluminium foil boat into which the slide containing the tissue section or cell suspension was placed and overlaid with PCR reaction mix and mineral oil.

Optimization of thermal conduction form the heating block to the sample must be achieved if successful amplification is to occur. One of the most critical parameters in PCR is the attainment of correct annealing and denaturation temperatures at the level of the tissue section, where the amplification process occurs. 'Thermal lag' (i.e. differences in temperature between the block face, the glass slide, and the PCR reaction mix at each temperature step of the reaction cycle) occurs, but this has not been adequately addressed by many authors. This phenomenon has been examined and a significant temperature differential between the sample and that of the block during the denaturation, primer annealing, and extension phases of the reaction has been found (1). In some cases this is of the order of 3–4°C, which in most cases contributes to total reaction failure, largely because initial denaturation temperatures are never achieved. Thermoconduction in most protocols has been maximized by filling unused air spaces with water or mineral oil, or by placing the glass slide in an aluminium foil boat. Alternatively, the heating block and glass slides can be covered with a plastic lid to optimize heat trapping and humidity.

Ideally, if using a standard thermocycler or heating block, a thermocouple should be used to assess the temperature of the slide, block, and sample (where possible) during each of the steps of the amplification process. Recently, machines have become available which offer in-built slide temperature calibration curves, which correct for thermal lag phenomena. Several commercially available machines, e.g. those from Perkin–Elmer, Hybaid, M.J. Research, Barnstead/Thermolyne and Coy Corporation correct for this problem, with specifically designed thermal blocks that optimize thermal conduction from the block face to the glass slide and thence to the tissue section.

The next sections detail the physics and calibration strategies of two of these newly developed machines and explore the evolution of *in situ* amplification technology and chemistry.

2. The GeneAmp® *In Situ* PCR System 1000

STEVE PICTON and DAVID HOWELLS

2.1 Introduction

There is no doubt that the major experimental technique to impact on the fields of molecular and developmental biology in the past decade has been PCR. From the first report of *in vitro* amplification of DNA (2) and the subsequent isolation and use of a thermostable DNA polymerase to allow automation of the PCR amplification process (3), the technique has been widely applied to areas of both fundamental and applied scientific research as diverse as molecular archaeology, the forensic sciences, clinical pathology, a variety of genome sequencing projects, and environmental monitoring. From the outset of the PCR process the dream of performing the amplification process *in situ* to enable amplification, detection, and localization of target nucleic acids in non-disrupted cells and tissues has excited both molecular pathologists and developmental biologists. For workers in these fields the technique would open up new degrees of sensitivity in the detection and localization of target sequences.

The first successful DNA amplification *in situ* was published by Haase and co-workers in 1990 (4), amplifying lentiviral DNA in cells and detecting the product by subsequent *in situ* hybridization. There followed reports of successful *in situ* amplification of human papilloma viruses in formalin-fixed tissues (5,6), amplification and detection of viral DNA, single-copy genes, and gene rearrangements in cell suspensions and cytospins (7), the detection of HIV in clinical samples (8), use of *in situ* PCR amplification to detect low copy number sequences in chromosome spreads (9), and identification of HIV-1 provirus in individual cells by coupled *in situ* amplification and flow cytometry (10). More recently the technique has provided the enabling technology to allow localization of Kaposi's sarcoma associated herpesvirus (KSHV) in Kaposi's sarcoma (11). Despite the increase in publications citing *in situ* PCR amplification, the routine use of the technique has remained until recently confined to a relatively small number of research groups. Other workers report encountering problems resulting in failure to amplify the target amplicon, near or complete loss of tissue morphology, or failure to detect and localize the amplicon following PCR.

2.1.1 PCR: a brief historical perspective

The rapid march of solution-phase PCR from a complex experimental process in the hands of a few dedicated scientists, the position only a decade ago, to a simple routine molecular tool arose from three main factors:

(a) The development of high quality automated instrumentation made it possible to achieve accurate and reproducible temperature conditions to ensure reliable, sensitive, and specific amplification.
(b) Cloning and recombinant expression of thermostable DNA-dependent DNA polymerases led to reduced price, increased availability, and high uniformity of thermostable enzymes such as AmpliTaq DNA-dependent DNA polymerase.
(c) The almost instant availability and reduced price of sequence-specific oligonucleotide primers allowed the researcher to design experiments, synthesize primers, and examine the experimental PCR results within a working week.

Towards reliable PCR amplification in situ

To date one of the major limitations in the reproducibility and portability of *in situ* PCR amplification, and thus its general acceptance as a valuable research tool, has been the unavailability of reliable instrumentation and enzymology dedicated to the experimental process. This section will deal with the work of Perkin–Elmer's Applied Biosystems Division to develop a dedicated instrument and sample containment system to automate and allow rapid, reproducible, and reliable amplification of nucleic acids in fixed cells or tissues on microscope slides and developments of Roche Molecular Systems to provide reagents specifically to address the needs of amplification *in situ*.

Why has the technique of *in situ* PCR amplification, an enabling technology for both amplifying and localizing target nucleic acid sequences to individual intact cells, taken so long for its use, reproducibility, and validity to be generally accepted by the scientific community? The technique of *in situ* PCR relies on the linking of four individually optimized experimental steps. Samples must be fixed and then permeabilized to allow access of PCR reagents. Following addition of reagents, PCR amplification is carried out and the products are then detected and visualized. Lack of accurate, reliable, and reproducible instrumentation for sample containment and *in situ* PCR amplification on microscope slides could prove to be the cause of failure even if all other variables were eliminated.

2.2 PCR amplification *in situ:* problems and pitfalls

We will focus here on issues relating to the reliability and reproducibility of the amplification process *in situ* and the suitability of hardware for carrying out PCR amplification on slide-immobilized, fixed, and permeabilized cells or tissues. The main criteria when examining the requirements for such instrumentation are:

(a) in order to minimize damage to tissues and ensure that amplification efficiency is maintained near to its maximum the sample target temperatures must be accurately met and maintained throughout the amplification process;

(b) to ensure that the experimental material and reagents reach the required set temperature, there must be no temperature gradient between the sample and the block;

(c) a reliable containment system is required so that evaporation of reaction components and drying out of the tissues during cycling are completely eliminated;

(d) an ability to set up reactions incorporating a 'hot start' would minimize mis-priming events and dimerization of oligonucleotides to increase further the specificity and sensitivity of the amplification process.

2.3 *In situ* PCR: the case for dedicated instrumentation and sample containment systems

Examination of the experimental variables that allow reliable PCR amplification *in situ* often overlooks a key factor, i.e. the accuracy, uniformity, and reliability of the equipment used. It is essential that the cycling method being used is both accurate and reproducible so that conditions achieved by all samples in one experiment can be exactly repeated later. Early papers reporting PCR amplification *in situ* often used a modified tube thermal cycler, hybridization ovens, or extensively modified laboratory equipment. Convection hybridization ovens, whilst providing for high throughput and temperature uniformity, are very slow when used to cycle samples between temperatures. One published protocol uses a single cycle time of 19 min (12). Lengthy incubation periods at elevated temperatures will obviously compromise preservation of tissue morphology and lead to problems when trying to define cellular localization of the amplified template. A frequent modification to existing thermal cyclers to perform *in situ* PCR is to place a mineral-oil-filled aluminium boat on to the cycler block and immerse the prepared reactions sealed on to microscope slides into the oil for the amplification steps. Such an approach, though sometimes successful, is subject to similar problems encountered using heated plate cyclers (flat blocks). Block temperature *per se* is not the critical factor for successful amplification: the actual temperature attained by the sample is of paramount importance. Using oil-filled boats, with poor and non-uniform thermal transfer between the cycler block and the experimental sample, the researcher is unlikely to achieve a robust, reliable, and repeatable technique. Flat-block cyclers are further susceptible to a large thermal gradient between the upper heated cycler surface and the upper surface of the slide. It is essential, therefore, that the temperature of the block surface is regularly calibrated against measured (observed) sample temperature on the surface of the slide; experiments with thermocouples demonstrate a difference of several degrees Celsius being common. The potential temperature differential between block and sample becomes greater at the higher temperatures critical for denaturation and will be dependent upon ambient laboratory temperature. In some cases the differential between the measured block temperature and that of the sample may be greater than 5°C and variable (13). In some cases the total failure of the technique may be explained by the failure of the sample to achieve a temperature at which the target nucleic acids are completely denatured. Clearly these approaches are unlikely to achieve the reliability, repeatability, and portability that such a technique requires.

Whilst not a scientific factor limiting the success of the technique, but undoubtedly affecting its overall popularity, *in situ* PCR was until recently an extremely messy and labour-intensive process. Sample and reagent containment and the prevention of reagent evaporation required the use of glass coverslips held in place on the microscope slide with nail varnish or silicone sealant. Alternatively people have created shallow wells around the sample with silicone adhesive prior to addition of reagents and then sealed the reaction receptacle with a coverslip. Unfortunately the pressure exerted by expansion of reagents during the repeated cycles of heating and cooling means that the coverslips rarely remain sealed in place (13) and thus, in order to halt evaporation of the reagents during amplification, the entire assembly had to be overlaid with mineral oil, a time-consuming procedure that introduces large thermal variability that may lead to failure of the sample to reach target temperature and further precludes daily use in high throughput situations such as molecular pathology laboratories.

2.4 A dedicated high-performance system for PCR amplification on glass slides

In order to address the issues of guaranteed sample temperature, run-to-run repeatability, and reagent containment as discussed above, a dedicated *in situ* thermal cycler and containment system, the GeneAmp *In Situ* PCR System 1000, has been developed by Perkin–Elmer's Applied Biosystems Division. Reagents suitable for *in situ* PCR amplification have been developed by Roche Molecular Systems to complement the Perkin–Elmer instrumentation.

2.4.1 The dedicated slide thermal cycler

The GeneAmp *In Situ* PCR System 1000 Thermal Cycler, based on the Perkin–Elmer DNA Thermal Cycler 480, contains an aluminium sample block consisting of ten parallel rows of vertical slots (see *Figure 1*). Each slot contains a pair of spring clamps at the front and back of the block that push a

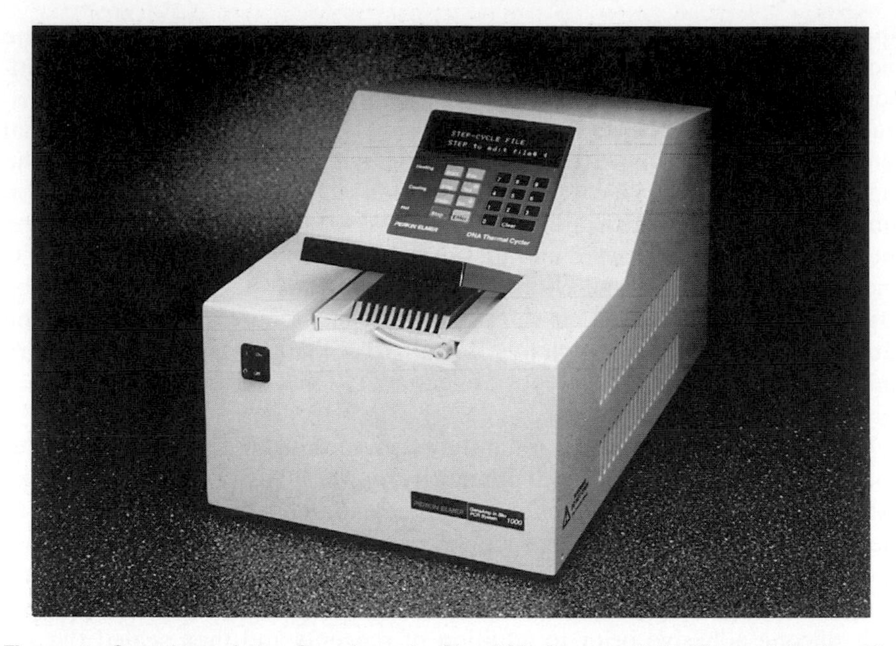

Figure 1. Overview of the GeneAmp *In Situ* PCR System 1000 Thermal Cycler. The aluminium block contains ten vertical slots each of which accepts a microscope slide with up to three individual AmpliCover containment assemblies.

single glass slide, with up to three containment systems, on to the left-hand side of the finned block surface, achieving intimate contact between the microscope slide and the block's finned surface, and thus ensuring a perfect thermal contact, no thermal gradient between block and sample, and thus reliable and reproducible target temperatures independent of ambient conditions (see *Figures 2* and *3*).

Since the slide and assembled containment system(s) are now held between two identically heated and cooled fins (see *Figure 2*) it is possible routinely to achieve highly precise and uniform thermal cycling of the experimental material (see *Figure 3*). Each slot accommodates a single glass slide on to which up to three separate samples and containment systems can be mounted, allowing up to 30 samples to be processed simultaneously in the thermal cycler. The thermal cycler uses embedded cartridge heaters and a self-contained circulating refrigeration system to achieve user-defined thermal cycling in the temperature range of 4–100°C. The high precision of block production, coupled with the block's compact design, provide the user with a static block uniformity of better than ±0.5°C (*Figure 4*). The aluminium block contains two solid-state feedback sensors. The cycler software enables the machine to run on calculated sample temperature since the thermal kinetics of the block and its thermal transfer to the slide, containment system, and

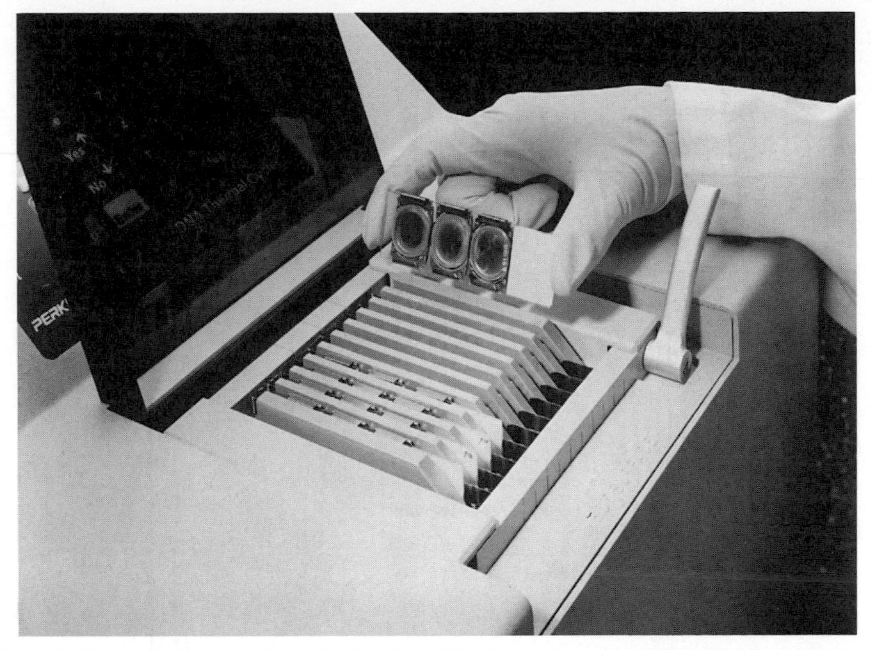

Figure 2. Instrument operation. An *in situ* PCR glass microscope slide, with up to three individual AmpliCover containment assemblies, is placed vertically into the block prior to PCR amplification. Each slide is held firmly against the finned block surface by spring clips operated via the lever visible in the foreground.

PCR reagents are all known (see *Figure 3*). This innovation removes the need for the user to calibrate sample temperature versus block temperature with external thermocouples. All instruments are factory-calibrated to standards traceable to NIST, thus ensuring total run uniformity between instruments, again a variable not previously addressed.

Thermal cycler programming facilities

The thermal cycler software contains four default programs (each of which can be accessed, edited, and re-saved). These enable the user to program indefinite holds, holds at a given temperature for a defined time, and cycling files either with or without variable rates of ramping between target temperatures. There are also two demonstration *in situ* PCR cycling programs. Storage memory facility allows saving of up to 93 user-defined programmes with up to 99 repeat cycles per program. Facilities exist for program-to-program linking, auto-segment extension facilities, and ramp time programming.

2.4.2 Integrated containment system for samples and reagents

To remove all experimental variation associated with non-uniform reaction volumes and increase the ease and speed with which the *in situ* PCR amplification procedure can be carried out, a novel reagent containment system has

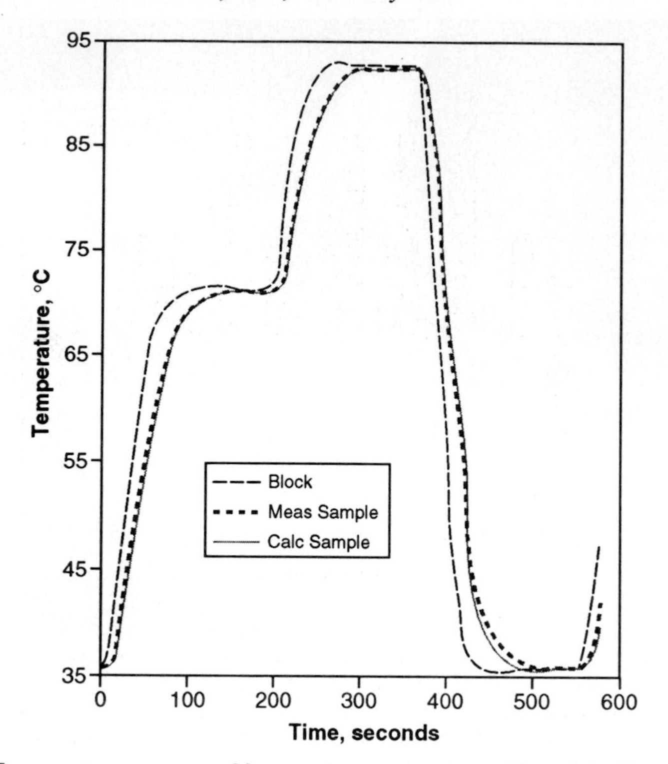

Figure 3. Temperature accuracy. Measured temperature profiles of the GeneAmp *In Situ* PCR System 1000 Thermal Cycler. Sample (slide) and block temperatures were measured with fine wire thermocouples. Calculated sample temperature as used by the PCR System 1000 was computed from measured block temperature assuming a pure exponential thermal time constant of 17.8 sec. The accuracy of the calculated sample temperature was validated by similarity to the observed sample temperature profile obtained with a thermocouple bonded on to a glass slide. Temperature cycle was 2 min at 72°C, 2 min at 94°C, 2 min at 35°C.

been developed. The sample and reagent containment system is integrated with the slide thermal cycler described above. Consisting of pre-silanized 1.2 mm thick *in situ* PCR microscope slides, batch-checked for uniform thermal conductivity, a concave silicone rubber gasket (the AmpliCover Disc), and a three-part stainless steel clip (the AmpliCover Clip) (*Figure 5*), the system enables previously fixed, sectioned, and permeabilized tissue sections attached to the slide surface to be flooded with PCR reagents and then be quickly sealed. The containment assembly provides a pressure-tight seal over the experimental sample and a standardized reaction volume whilst removing the need for nail varnish, brushes, adhesives, and mineral-oil overlays. Rapid assembly of the two-part containment system on to the slides is achieved using the Assembly Tool (*Figure 6*). This has a heated aluminium assembly platen,

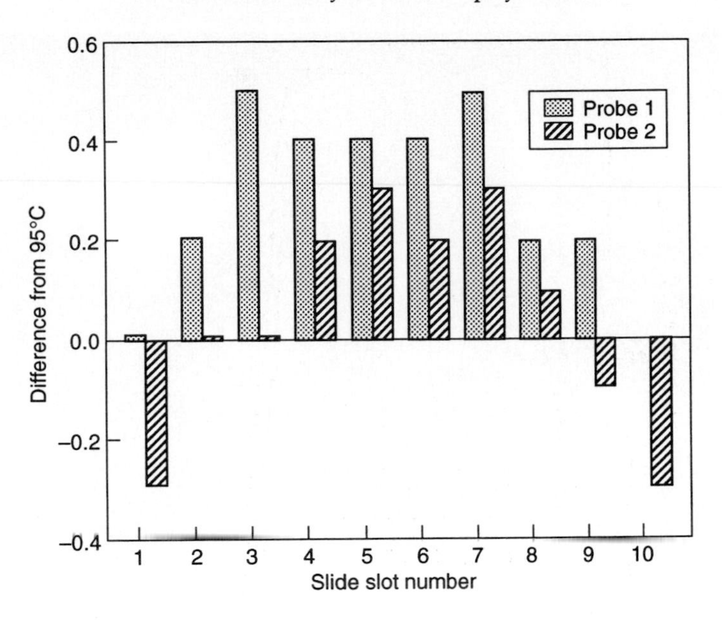

Figure 4. Block accuracy and uniformity on the GeneAmp *In Situ* PCR System 1000. Steady-state thermal uniformity of the block at a programmed 95°C was measured at two positions with calibrated platinum resistance chips bonded to an *in situ* PCR glass slide. Measured block uniformity from slot 1 to 10 and from front to back of individual slots was better than ±0.5°C.

factory set and calibrated at 70°C. A temperature sensor and flashing LED indicate to the user when the platen has reached the correct temperature, to enable the user to carry out 'hot-start' *in situ* PCR. The ability to set up reactions under 'hot-start' conditions ensures that non-specific amplification resulting from template–primer complexes formed below primer T_m and primer dimerization are eliminated. This increases the specificity, sensitivity, and reproducibility of the PCR amplification process.

The development of this cycler and an integrated containment and assembly system removes all experimental variables associated with thermal gradients, variable reaction volumes, and failure of the sample to reach the programmed temperature. The system also eliminates the variability—in some cases complete failure—that results from evaporation and drying out of the experimental tissue and reagents during amplification. Block variability and thermal gradients between sample and block are almost completely eliminated by virtue of the vertical block design whilst the thermal cycler's independent heating and cooling systems allow for high temperature precision, cycle-to-cycle reproducibility, and ability to operate to high precision independent of ambient temperature.

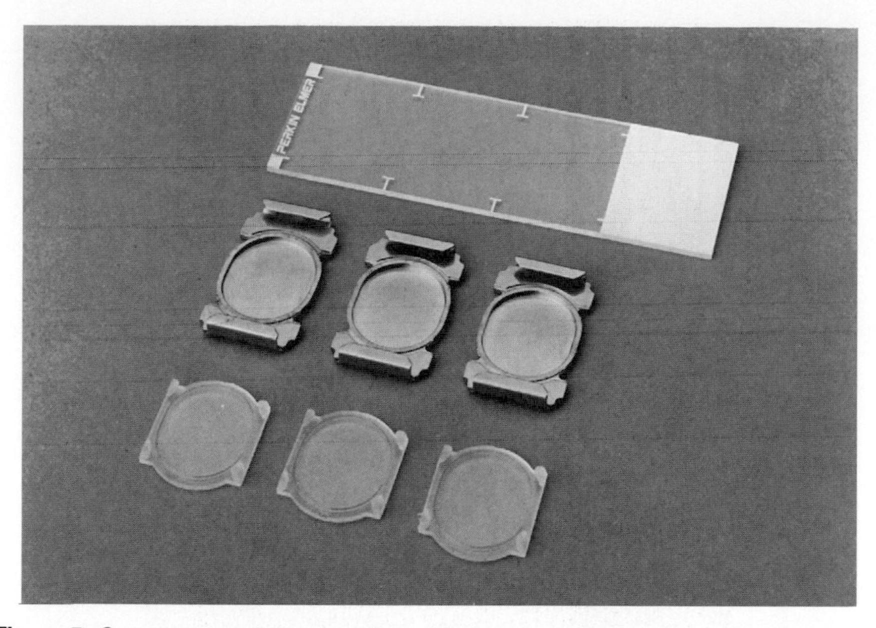

Figure 5. Components of the AmpliCover Containment System. In order to provide a liquid-tight seal of PCR reagents over the samples immobilized on the slide, a two-part containment system is used. The slide, stainless steel AmpliCover Clips and silicone rubber AmpliCover Discs are shown ready for assembly.

2.5 Dedicated and qualified reagent system for *in situ* PCR amplification

Optimal reaction component conditions for PCR amplification *in situ* vary from those commonly used for solution-phase PCR. Tissues, cell smears, cytospins, etc. are mounted on glass microscope slides. These slides are generally pre-treated with silane to ensure adhesion of the experimental material to the slide surface. This highly charged slide surface along with the cellular material undoubtedly leads to sequestration of both the thermostable AmpliTaq DNA polymerase and its essential co-factor, magnesium. Clearly the surface area:volume ratio of a reaction performed on a microscope slide will differ from that achieved in a tube and will further lead to potential reduction in amplification efficiency due to loss of components binding to the reaction vessel surface (for detailed discussion of these factors see ref. 13). To address these issues it is essential that magnesium levels are determined empirically for each tissue. To ensure that the DNA polymerase is present in excess, more units of polymerase must be used in an *in situ* reaction. Our partners in research, Roche Molecular Systems, have developed a novel formulation of AmpliTaq specifically formulated for use *in situ*. The GeneAmp *In Situ* PCR Core Kit contains all reagents, excluding primers, water, and label, needed to carry out

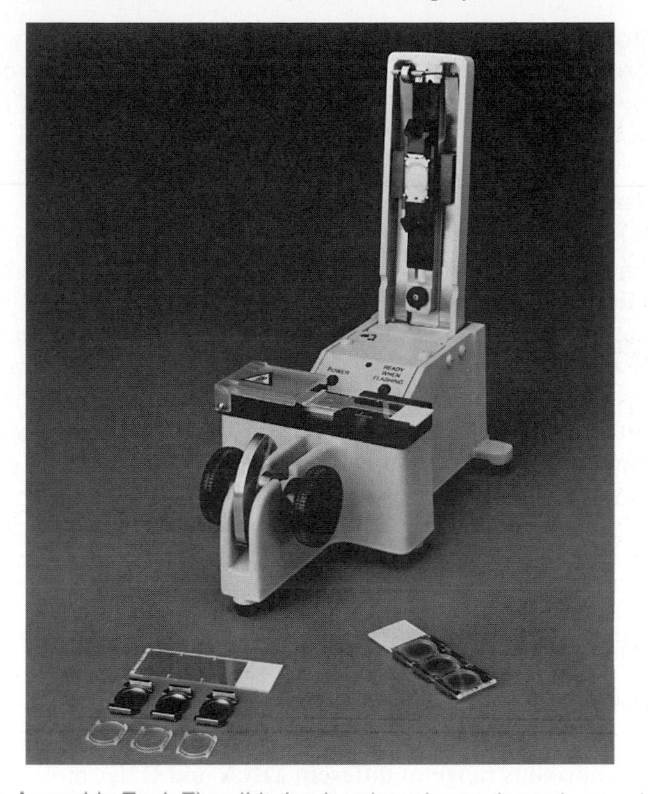

Figure 6. The Assembly Tool. The slide is placed on the pre-heated assembly platen in the centre of the Tool as shown. The AmpliCover Clip, with Disc inserted, is placed into the handle of the Assembly Tool and held in place by a magnet. The handle is drawn down over the slide and a concentric cam operated by the black handles engages with the lid and draws the AmpliCover assembly on to the surface of the slide, forcing out air and creating a liquid-tight seal. Two push-operated levers on the upper surface of the lid engage the stainless steel clips on to the slide to create a reaction containment system as shown (below, right). After cycling the clips are simply removed for re-use using a Disassembly Tool (not shown). AmpliCover Clips, Discs, and slides are shown individually (below, left).

in situ PCR amplification. Consisting of AmpliTaq DNA Polymerase IS (20 U/μl) and 10× detergent-free buffer system, with separate magnesium and dNTPs, these components are designed to integrate with the GeneAmp *In Situ* PCR System 1000 to ensure reliable and robust amplification. The high unit concentration of AmpliTaq DNA Polymerase IS allows addition of more units of enzyme without the carry-over of increased levels of glycerol present in the enzyme storage buffer. The presence of high concentrations of detergents, commonly used in thermostable enzyme preparations, such as Nonidet P-40 and Tween during the cycling process, may lead to unacceptable disruption of sample morphology. AmpliTaq DNA Polymerase IS is supplied with a detergent-free 10× buffer system and operates optimally in the presence of a proprietary detergent that is contained only in the enzyme storage buffer. Separate magnesium allows for empirical determination of magnesium optimum.

2.6 Summary

In a recent review of the techniques of *in situ* PCR amplification (14) it was suggested that 'the need for a specifically dedicated instrument to perform these techniques is now being addressed'; the authors further state that, 'standardization of instrumentation and simplification of the technique are required in the near future, if the technique is to receive universal acceptance.' The release of the Perkin–Elmer GeneAmp *In Situ* PCR System 1000 and associated consumables overcomes the problems associated with the technique and provides the researcher with the first instrument, containment, and reagent system dedicated to performing *in situ* PCR.

3. Automation and the Omnislide System

JACQUELINE A. STARLING

3.1 Introduction

Since its description in 1969 (15,16), *in situ* hybridization (ISH) has been used to detect an enormous range of different DNA and RNA molecules. While it is an extremely powerful technique, it does have a lower limit for target sequence detection which restricts its usefulness within some areas of research. The reliable threshold levels for standard ISH detection of a target sequence within a single cell are approximately 10–20 copies for mRNA and 10 copies for viral DNA (reviewed in ref. 17; see also Chapter 3).

Sensitivity limitations were overcome in 1990 with the arrival of PCR ISH. Here, an amplification step is incorporated into the ISH protocol to increase the copy number of the target sequence to a level where it can be readily detected. It has since been shown that PCR ISH allows the detection of a single target copy of a specific sequence per cell; for example, single-copy genes (7) and one or two copies of viral DNA (18,19), have been successfully amplified and localized.

Traditionally ISH techniques have been performed using a variety of equipment already available within the laboratory. While this 'makeshift' approach has proved to be sufficient, and has seen the generation of many scientific publications, there are a number of significant drawbacks, namely slide handling limitations and inaccurate, unreliable temperature control. With the arrival of PCR ISH came the additional need for accurate thermal cycling of samples on slides and a way to prevent the sample drying through the constantly changing, elevated temperatures. Neither of these additional requirements was possible using the standard equipment available, highlighting the need for a thermal cycler designed specifically for slides.

In addressing the need for a slide thermal cycler we, at Hybaid, decided to look at the requirements of the whole PCR ISH protocol. The ultimate goal of any *in situ* protocol is to obtain specific, reproducible results with low levels of background, strong specific signal intensity, and optimal sample morphology. In order to meet these criteria, the instrumentation used must provide accurate sample temperature control, humidity maintenance, and a means of handling large numbers of slides. Analysis of the 'makeshift' methods for performing *in situ* techniques revealed that each of these factors could be improved by semi-automating the way in which the protocols were performed. Hence, our aim was to develop a system that catered for the permeabilization, target sequence amplification, hybridization, detection, and washing stages of PCR ISH. In doing this we not only answered the need for a slide thermal cycler but also dedicated *in situ* instrumentation that would improve both the quality and reproducibility of results generated by all *in situ* protocols (ISH (see Chapter 3), primed *in situ* labelling (PRINS, see Chapter 6), PCR ISH (see Chapters 4, 5, 7, and 8), and immunohistochemistry)).

3.2 Factors and problems to be addressed in order to semi-automate PCR ISH

3.2.1 Slide handling

During the course of a PCR ISH protocol, the slides bearing the sample are passed through a series of wash steps requiring their vertical insertion into wash buffer in Coplin jars or cuboidal baths. The wash steps are interspersed by incubations where the slides are individually blotted and placed flat on to the base of a moist chamber and small volumes of reagents are pipetted on to the sample (e.g. enzymes, hybridization buffers, or antibodies). The constant transfer of slides between the vertical washing steps and horizontal incubations is not only extremely time-consuming, particularly if 20 slides or more are being processed simultaneously, but also increases the possibility of slide breakage or mechanical damage to the sample. Additional problems that are encountered include samples drying out while being transferred between vertical and horizontal stages and a wide variation in the time for which individual samples are exposed to each reagent.

3.2.2 Temperature control

For incubation steps above ambient temperature a variety of ovens and waterbaths are used. For horizontal incubation steps the slides are placed into a moist, sealed chamber and either floated in a waterbath or left in an oven. If a DNA target is to be localized it is often denatured after addition of the probe, for example, at 95°C. This is then followed by the transfer of samples to the desired hybridization temperature. This step necessitates the use of two

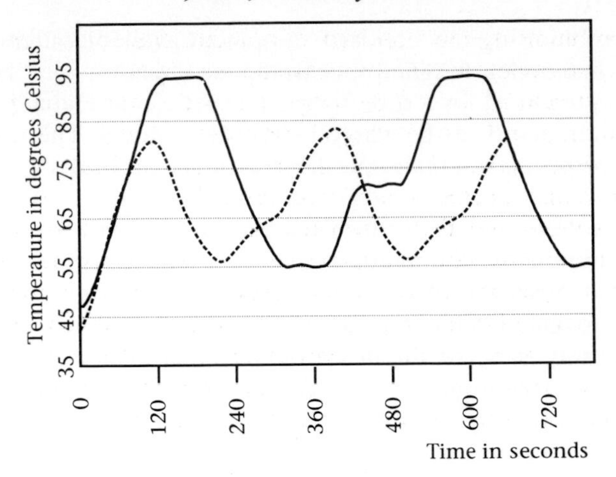

Figure 7. Improvized thermal cycling of samples on slides. The trace shows the different sample temperatures achieved using simulated slide control on Hybaid's flat-block thermal cyclers (solid line) and when a slide is placed within an aluminium boat on a conventional tube thermal cycler (broken line). The sample on the slide within the aluminium boat never attains the set temperatures. Programmed cycles: 94°C for 1 min, 55°C for 1 min, 72°C for 1 min.

waterbaths or ovens or a long waiting period while one oven or waterbath cools to the desired temperature. Additionally, for short incubations, like the denaturation or permeabilization steps, an approximation of the time taken for the sample within the box to reach the set temperature must be added to the incubation. For washing steps Coplin jars or cuboidal baths, containing the wash buffers, are placed inside ovens or waterbaths and the temperature of the wash buffer allowed to equilibrate before the wash steps are commenced.

When performing the amplification step of a PCR ISH protocol the samples need to be passed rapidly and accurately between the denaturation, annealing, and extension temperatures for a variable number of cycles. The use of ovens and waterbaths set at different temperatures is impractical and hence rarely used. Instead researchers have attempted to perform amplification protocols by placing their slides in hand-made aluminium foil boats on top of a conventional thermal cycler heating block. These thermal cycler blocks were designed to accommodate 0.5 ml plastic tubes which would make direct contact with the heating block. Placing a flat aluminium foil sheet on the surface of a heating block with holes to hold tubes gave inefficient heat transfer to the slides to the extent that they never reached the set temperature (see *Figure 7*); there was even a lack of temperature uniformity across individual slides and, as a consequence, results were often unreliable.

3.2.3 Humidity control

During both PCR ISH and ISH incubation stages, it is very important that the sample remains immersed in a solution of reagents containing the appropriate buffer, probe, enzymes, etc., and preventing the evaporation of this solution is crucial to the success of the experiment. Scientists performing hybridization, permeabilization, and detection steps simply cover the sample with reagents and a coverslip and place the slide flat into a plastic box containing paper tissue soaked in an appropriate buffer solution. The sealed box is then placed in an oven or waterbath if above-ambient temperatures are required. Solution from the tissue paper evaporates, creating a humid environment which, in turn, reduces the evaporation of solution from the surface of the sample. For PCR ISH protocols where samples are subjected to rapidly changing temperatures the creation of a humid environment using the 'box' method is not possible. Instead the complete, sealed enclosure of sample and reaction solution is essential for sample and reagent integrity.

3.2.4 Summary

At every stage of both PCR ISH and ISH protocols there is the potential for the introduction of factors which could severely alter the outcome of the experiment. These factors can be a reflection of the limitation of the instrumentation being used for a purpose for which it was not designed, inadequate preparation of samples, or, simply, errors introduced by various monotonous handling routines. In these, and other cases, semi-automation would clearly address the need to limit these factors to a minimum such that the sensitivity and fidelity of the results obtained could be reproducible not only in the hands of the laboratory generating the original results but also in the hands of scientists, with limited experience, who wish to adopt these protocols as a research tool.

3.3 System design: theoretical and practical considerations

In designing a system to simplify the technically challenging aspects of PCR ISH experimentation, Hybaid has taken a 'whole systems' approach. Hence, every stage of a typical PCR ISH protocol, from pre-treatment, through thermal cycling, to detection of signals, has been analysed carefully and a suitable solution, in the form of a specific feature, incorporated as part of a complete, flexible system. Furthermore, the demands associated with optimizing these techniques within a laboratory have been considered in the design of these systems.

3.3.1 Absolute temperature control

It has long been recognized that accurate control of temperature is essential for reproducibility, low background, maintenance of sample morphology, optimal signal intensity, and specificity of *in situ* results. Particular attention must therefore be paid to the achievement and maintenance of precise sample temperatures, through single-step incubations and thermal cycling, when

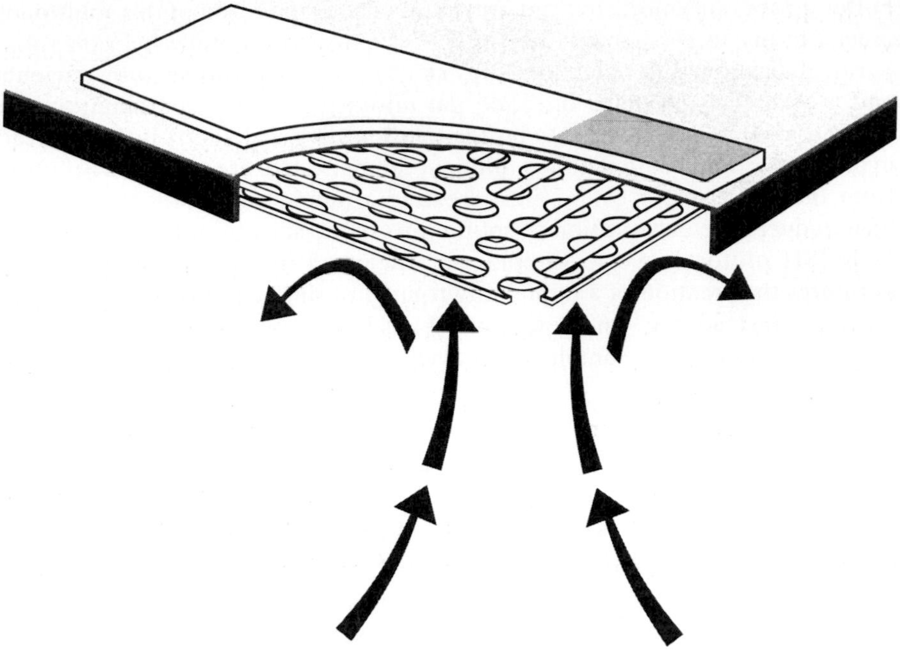

Figure 8. Slide thermal cycler flat-block heating and cooling set-up. The flat block of the slide thermal cycler is heated by means of a wire heating element wound into a copper frame to a specific, pre-determined pattern positioned in direct contact beneath the block's surface. Rapid cooling is achieved by powerful fans situated beneath the block. The air current pattern they generate is indicated by the arrows shown.

designing any absolute temperature control system.

After analysing many slide heating methods such as air convection, water conduction, and microwave technology, we found that the most accurate and rapid heating is achieved when the slide is placed directly in contact with the surface of a flat block. The flat block is cast in a metal with good conductance properties, for example, aluminium or silver, and is heated by means of a wire heating element positioned in direct contact beneath the block's surface. This element is wound into a copper frame to a specific, pre-determined pattern and optimized to transfer heat to the blocks rapidly while maintaining uniformity of temperature across the block. Rapid cooling is achieved by powerful fans situated beneath the block. These generate air currents which flow past the underside of the block and exit via air vents shaped to duct away the heat exhaust (see *Figure 8*). The thickness of the block is finely tuned to give optimal mechanical stength, while the mass of the block is not so great that it limits the speed of temperature transitions.

Temperature control software

To maintain temperature accuracy the rate of heating and cooling of the flat-block is regulated by temperature control software. Thermistors located along the underside of the block constantly monitor its temperature and feed this information back to the control software which accordingly heats or cools the block. Two methods of temperature control are commonly available: block control and simulated slide control.

Block control

In the block control method the block on which the slides rest will heat up to the desired temperature and remain at that temperature for the programmed time span. This method of control does not take into account the temperature of the sample on the slide which lags behind as the block is heated; this lag reflects the time taken for the heat to pass through the glass slide and heat the sample. Eventually the sample will reach the set temperature. However, as the control software is designed to start timing from the moment the block reaches this point, it is not an indication of how long the sample spends at this temperature (see *Figure 9*). This is particularly relevant for thermal cycling protocols where it is critical that the sample reaches the desired temperature

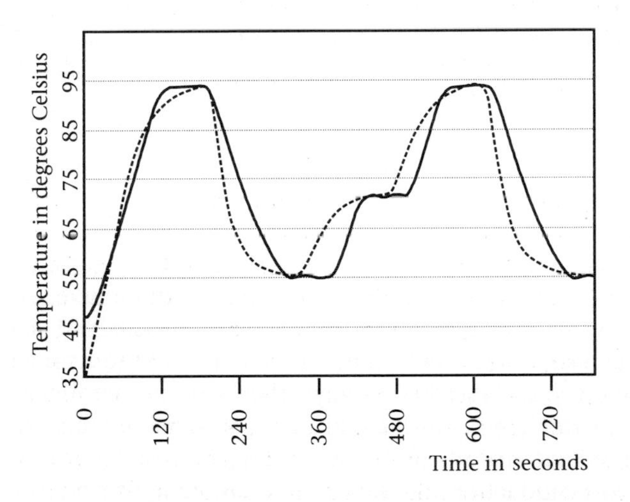

Figure 9. Comparison between block and simulated slide control software. The trace shows the different sample temperatures achieved on a slide using simulated slide control on Hybaid's flat-block thermal cyclers (solid line) and using block control on another dedicated slide thermal cycler (broken line). As can be seen, when using block control the samples do not reach the set temperature until the end of the set time period in contrast to simulated slide control. Programmed cycles: 94°C for 1 min, 55°C for 1 min, 72°C for 1 min.

for a specific time interval. For this reason it is recommended that simulated slide control is used instead of block control.

Simulated slide control

Simulated slide control software takes into account the time delay between the block and sample reaching the desired temperature. The thermal inertia of the slide adds to that of the block so the software needs to compensate for the lag in temperature caused by the additional mass. This is achieved by a software algorithm that controls the temperature of the block such that the sample always attains the desired temperature rapidly. At the point where the sample achieves the set temperature the software will hold that temperature for the set period of time before proceeding to the next thermal stage (see *Figure 9*). This timing is particularly important for the optimization of annealing temperatures in PCR ISH protocols and the denaturation steps needed for all *in situ* protocols where DNA is the target. In addition, this degree of sample temperature control aids the reproducibility of results from experiment to experiment. Not only is this important in academic research applications but also it may determine the feasibility of PCR ISH being adopted in clinical diagnostics.

Calibration factor

Another variable that can lead to erroneous results is the thickness of the glass slides used in *in situ* experiments. If simulated slide control is to work effectively, compensation must be made for this variable. Most slides used in *in situ* experiments are 0.7–1.3 mm thick. A standard, single calibration factor built into the slide control software has proved sufficient to be used with this range of slide depth. In practice thicker slides are not used but if, for example, the researcher wished to perform an incubation on a coverslip, the calibration factor would need to be lowered to compensate for the much lower mass of glass that needs to be heated.

Uniformity

A major benefit of designing a flat-block system is that accurate control over dynamic temperature uniformity can be readily achieved and means that regardless of a slide's position on the flat block it will remain within the same tight temperature specification as any other slide on the block. Hence, all of the samples within a batch will experience the same heating and cooling conditions throughout a protocol. Some systems hold slides stacked in a mantle and rely upon conduction of heat from a source at the base of the block to heat the samples above. However, temperature gradients are generated across the stack during thermal cycling which adversely affect the conditions experienced by all of the slides.

Using a flat-block set-up, dynamic uniformity of better than ±1°C can be generated (see *Figure 10*). This is actively monitored by thermistors placed on

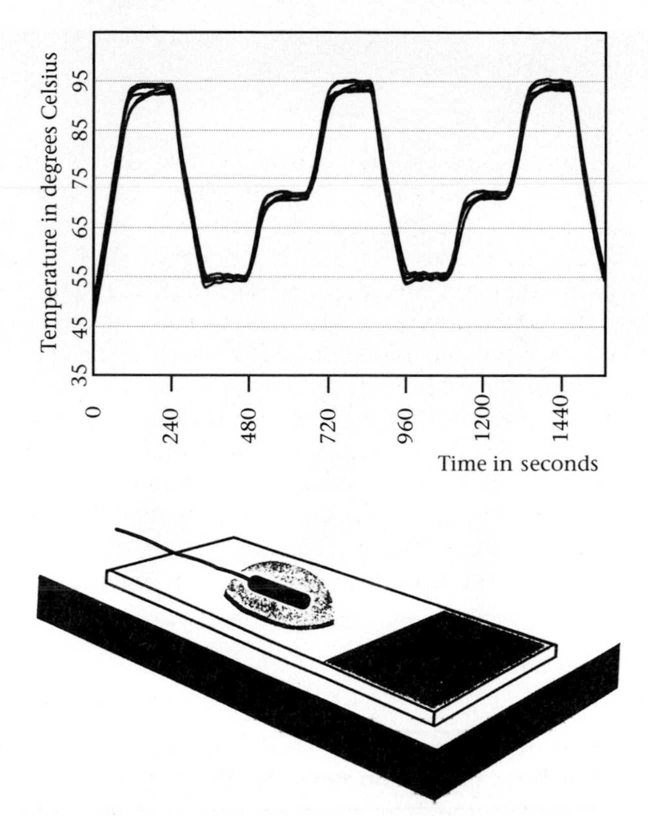

Figure 10. Dynamic temperature uniformity across the flat block of a slide thermal cycler. (a) The trace shows dynamic uniformity measured at six widely spaced positions across the OmniSlide dual flat blocks. Programmed cycles: 94°C for 2 min, 55°C for 2 min, 72°C for 2 min. (b) The actual sample temperatures achieved on a slide when a dynamic uniformity trace is generated are measured by placing thermistor probes, attached to glass slides, across the flat-block surface and monitoring the temperatures achieved during thermal cycling.

the underside of the block which feed temperature information back to the controlling software. Any alterations to heating or cooling are then made by this software. The size of the flat block will determine the number and spacing of the thermistors: a small four-slide block requires a single thermistor while a ten-slide block requires two thermistors positioned at opposite ends.

3.3.2 Slide handling

The number of slides and their method of handling at different stages of an *in situ* protocol can also determine the physical shape of the thermal cycling instrumentation. A reduction in the manual handling of slides is highly desirable especially where large numbers are involved. Traditional slide racks hold a variable number of slides in a vertical position and are used for washing

Figure 11. The Hybaid OmniSlide Slide Rack

steps where slides are transferred between reagents contained in cuboidal troughs. For low-volume incubation steps the slides are individually removed and laid flat. Designing a system around a rack that held slides vertically would be impractical for the temperature incubation stages. Additionally researchers need access to the surface of the slide during different stages of the protocol. For these reasons a rack has been developed that can hold up to 20 slides vertically for large-volume washing steps and horizontally for low-volume incubation steps (see *Figure 11*). During these stages the slides never need to be removed from the rack, thus minimizing the risk of physical damage to the samples and also saving time in handling slides during transfer from wash solutions.

3.3.3 Humidity control

The incorporation of a humid environment as an integral part of *in situ* instrumentation requires a dual approach. For prolonged static temperature incubations, in the range of room temperature to 55°C, sufficient humidity can be generated from a buffer solution contained within channels surrounding the heated flat blocks. For thermal cycling protocols this method of humidity control cannot prevent sample drying and it is necessary to seal a coverslip over the sample and reaction solution. To date this has been achieved, with a varying degree of success, using a number of materials of diverse composition; the most successful being rubber cement and nail varnish.

Figure 12. The Hybaid SureSeal.

To address the need for improved humidity control during thermal cycling protocols a specific slide-sealing device has been developed. Such a device must be flexible enough to allow the user to generate an impervious seal but also needs to be easily and completely removed before subsequent steps are performed. Different sizes of the device should be available to cater for the range of sample sizes used in PCR ISH and the volume of reagents required on the surface of the sample needs to be minimal to reduce the cost of each experiment. It is very important that the device is compatible with standard size glass slides, particularly where the use of archival material is necessary. Archival material is often pre-mounted on standard 1 mm thick glass slides and cannot be transferred to a slide and sealing system dedicated to any particular instrumentation. Where large numbers of slides are being handled the cost of sealing them with expensive custom-made devices can be a deterrent and the user may revert to using glues and varnishes which are unreliable. So the ideal device for sealing would be a cost-effective option that offered tangible benefits.

The Hybaid SureSeal slide-sealing device (*Figure 12*) addresses the above requirements. It is simply a plastic frame, with adhesive on both sides, that is placed around a sample on a slide. Reaction mix is then either pipetted on to the SureSeal coverslip and the two pressed together or it is pipetted on to the sample and the coverslip laid across. The adhesive ensures an air-tight seal and prevents sample drying through prolonged thermal cycling. Once thermal cycling is complete the Hybaid SureSeal is peeled off the slide in one piece.

Unlike other commercially available consumables, Hybaid SureSeals vary in size to enclose a 1 cm^2, 2.4 cm^2, or 4.75 cm^2 sample area to take into account the different sample sizes used in *in situ* protocols; when smaller samples are used up to three samples can be processed on the same slide, allowing a test sample and controls to be processed simultaneously. The volume of reaction mix needed for each frame has been kept to a minimum, i.e. 20 μl for 1 cm^2, 50 μl for 2.4 cm^2, and 100 μl for 4.75 cm^2, to reduce the cost of performing PCR ISH protocols.

3.4 Hybaid's *In Situ* Systems

3.4.1 TouchDown *In Situ*

Frequently, the possible use of PCR ISH in the design of a research project has meant that a pilot study into the potential of the protocol must be assessed before committing further funds and resources. (This reserved approach has undoubtedly been influenced by the lack of reproducible data published and also the novel nature of the technique.) TouchDown *In Situ* is a thermal cycler system that offers a four-slide capacity flat-block option as well as other regular block formats to address this level of commitment (see *Figure 13*).

The idea that a PCR protocol must first be optimized in solution and then

Figure 13. TouchDown *In Situ*. The TouchDown thermal cycler is shown with a 0.2 ml block control module and a 0.2 ml and four-slide capacity flat-block mixed satellite.

transferred to the slide format is central to the design of this system. Unlike interchangeable block formats found in other commercially available systems, TouchDown *In Situ* offers fixed *in situ* blocks available with both the control and satellite modules. This fixed block approach was adopted to answer the problems associated with interchangeability such as recalibration requirements, wear and tear of electrical connections, mechanical damage to the surface of the blocks, etc., which are frustrating and can hamper research efforts.

As a modular system TouchDown *In Situ* allows the independent programming, and operation, of up to three blocks simultaneously. Hence, if the laboratory, having attained success with a single flat block, should need to increase sample throughput, the system can be expanded to incorporate three *in situ* flat blocks with a total handling capacity of 12 slides. The triple-block TouchDown *In Situ* system would also aid the rapid optimization of new *in situ* protocols allowing different hybridization or thermal cycling conditions to be analysed simultaneously. In keeping with a 'whole system' approach, TouchDown *In Situ* can also be combined with the slide handling, washing, and signal detection elements of the larger capacity OmniSlide *In Situ* System.

3.4.2 OmniSlide *In Situ* System

The 'whole system' design approach to *in situ* instrumentation has led to the amalgamation of suitable solutions, engineered to satisfy the requirements of all *in situ* techniques. At the time of writing, OmniSlide is the only system that offers a semi-automated modular solution for every stage of an *in situ* protocol (see *Figure 14*).

The system comprises four separate modules:

- the OmniSlide Thermal Cycler, for hybridization, incubation, and thermal cycling
- the Ambient Wash System, for room temperature washing steps
- the Heated Wash Module, for above-ambient washing steps
- the Closed Humidity Chamber, for signal detection steps or room temperature incubations.

At the heart of the OmniSlide System is the Slide Rack (*Figure 11*), which allows up to 20 slides to be held and processed simultaneously. A 20-slide capacity was considered to be necessary to accommodate both the test slides and the numerous controls needed when performing PCR ISH. Once loaded into the Slide Rack, the slides do not need to be removed at any stage during an *in situ* protocol; the rack and samples are simply passed between each module of the system. Removal of the slides is only necessary when the results of the experiment are viewed under a microscope. Use of the Slide Rack reduces 'hands-on' time and, hence, handling errors and has been shown to improve results and reduce background.

The OmniSlide Thermal Cycler's dual flat blocks employ simulated slide

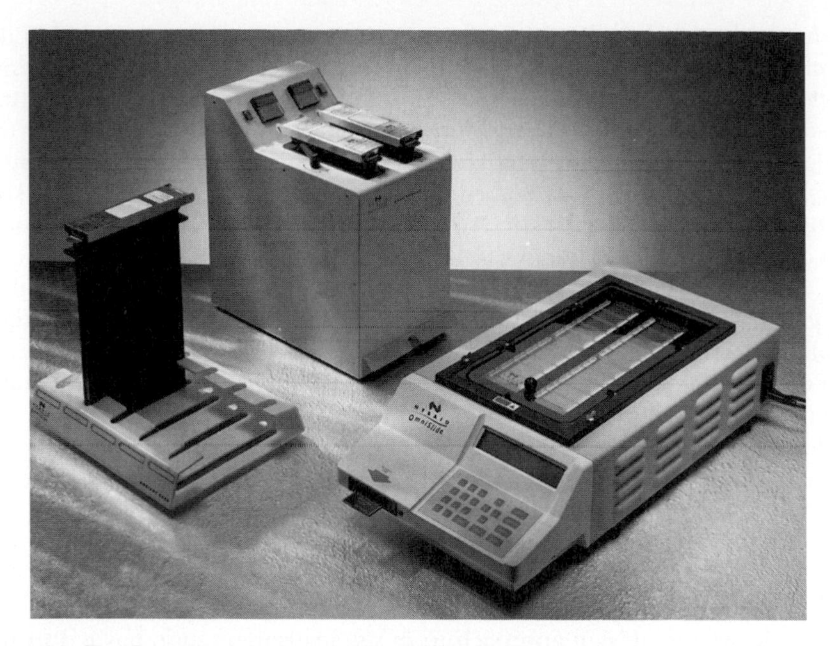

Figure 14. The OmniSlide System. The OmniSlide Thermal Cycler, Heated Wash Module, and Ambient Wash System are shown.

control to ensure that the required sample temperatures are achieved during the incubation, hybridization, and thermal cycling steps of an *in situ* protocol. The Slide Rack sits over the dual flat blocks, allowing the slides to rest in absolute contact with the heated surface, ensuring that the whole sample reaches the exact temperature required. To preserve sample integrity through both short and prolonged single temperature incubations a humidity chamber surrounds and encloses the flat blocks, Slide Rack, and samples. For thermal cycling steps where the slides must be sealed, both OmniSlide and Touch-Down can be used as 'open systems', i.e. researchers can use whichever slide format and sealing device best fulfils the demands of their protocol.

For all the washing steps, the Slide Rack is inserted into a Wash Sleeve held either at room temperature in the Ambient Wash System or at elevated temperatures in the Wash Module. The Wash Module contains two independently heated chambers designed to accommodate and uniformly heat the Wash Sleeves. It allows washes to be performed simultaneously and at two different temperatures from 5°C above ambient to 70°C.

The Light-Proof Humidity Chamber has been developed for the final signal detection step of an *in situ* protocol. Most *in situ* protocols using non-radioactive labels have a light-sensitive signal detection step. An example of this is the alkaline phosphatase NBT-BCIP (Nitroblue Tetrazolium/5-bromo-4-chloro-3-indolylphosphate) colour detection step of some protocols that use digoxigenin-labelled probes. This needs to be performed in a humid environment, in the dark. The fluorophores used for the techniques of fluorescence ISH (FISH) and PRINS also need to be kept out of natural light prior to excitation with the appropriate wavelength of light.

In summary, the OmniSlide System can be used for all applications of ISH, PRINS, PCR ISH, and immunohistochemistry. Its ability to process 20 slides simultaneously makes it suitable for the high-throughput laboratories and PCR ISH where the requirement for ease of sample handling is essential. The absolute control of sample temperature and humidity throughout any procedure delivers unparalleled, reproducible results.

3.5 Results achieved

Hybaid's *In Situ* Systems have been used successfully in the development and optimization of ISH (20,21), PRINS (9,22–26), PCR ISH (27–35), and immunohistochemistry protocols. They have been adopted by laboratories with a wide range of *in situ* experience from complete novices to laboratories that already have established protocols developed using a range of makeshift ovens, waterbaths, and slide racks.

3.5.1 Makeshift instrumentation versus the OmniSlide System

In designing the *In Situ* Systems Hybaid has addressed the need for improved temperature control, slide handling, and humidity maintenance in instrumentation used for *in situ* techniques. To assess the degree to which this has been achieved its performance was compared with that of makeshift instrumentation in ISH comparing results obtained using Amersham's DNA and RNA Colour Kits (RPN 3200, 3300) to detect pro-opiomelanocortin mRNA in rat pituitary tissue sections. For both the DNA and RNA Colour Kits, the results obtained from the OmniSlide System were superior in terms of signal intensity and levels of background (see *Figure 15*). This improvement was primarily due to the accurate sample temperature control achieved during the hybridization and stringency washing steps of the protocol. For the hybridization step, carried out at 42°C and 55°C for the DNA and RNA probes respectively, the improved temperature control generated a higher level of probe hybridization which was reflected in significantly shortened signal detection steps. In the case of the RNA Colour Kit, signal was easily visible after 2–3 h in contrast to the standard overnight incubation. The improved temperature control at both the hybridization and stringency washing steps reduced the levels of background seen for both probes. For the RNA probe this allowed information that was originally masked to be seen (see *Figure 15d*). The reduction in background could also be attributed to the minimal slide handling needed when the OmniSlide Slide Rack was used. As a whole the OmniSlide System not only improved the results seen but also improved their reproducibility from run to run.

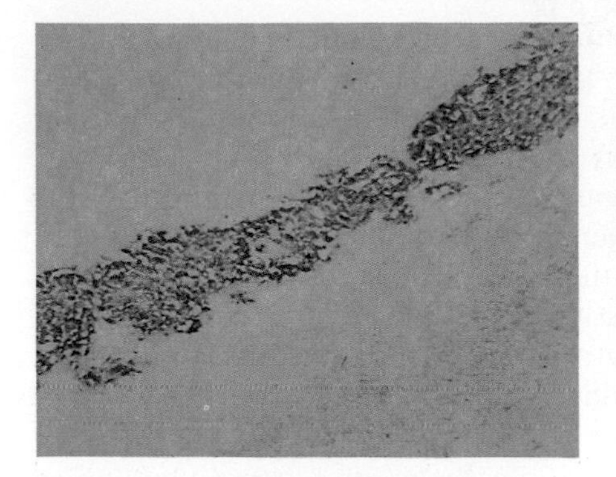

(a)

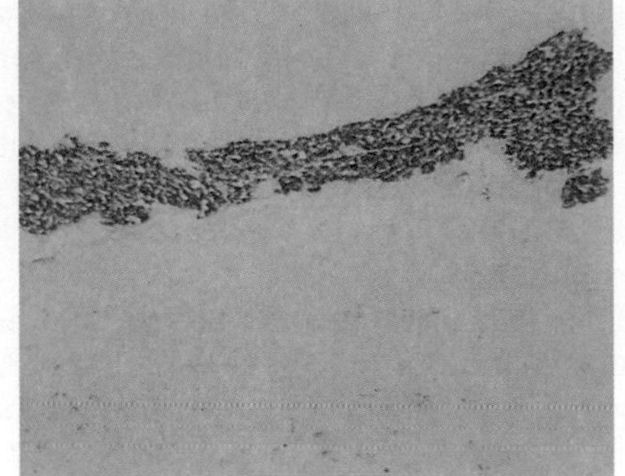

(b)

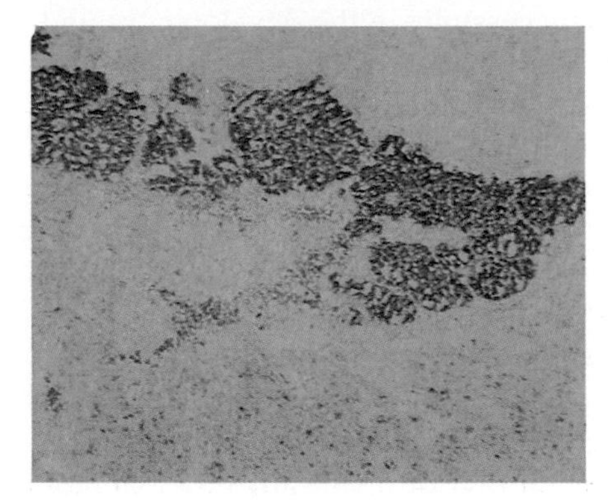

(c)

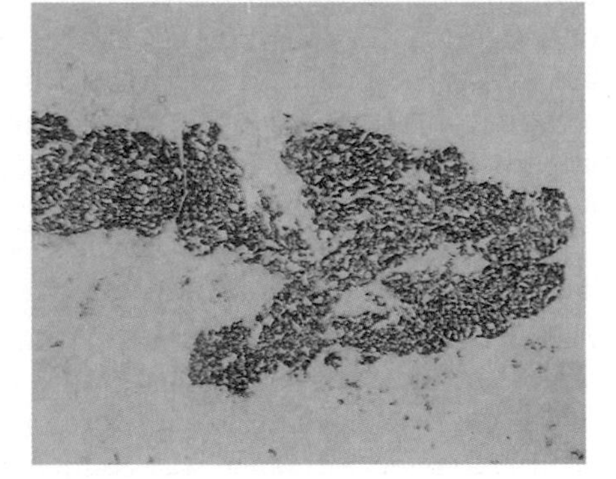

(d)

Figure 15. Comparison of ISH results obtained on conventional instrumentation and on the OmniSlide System. The Amersham DNA and RNA Colour Kits were used to compare the quality and intensity of signal obtained after ISH was performed to detect rat pro-opiomelanocortin (POMC) mRNA in rat pituitary tissue sections using either conventional instrumentation (a) and (c) or the OmniSlide System (b) and (d). The OmniSlide System gave better results with a much stronger signal intensity and improved levels of background (b) and (d). (a) Detection of POMC mRNA using the Amersham DNA Colour Kit and conventional instrumentation. (b) Detection of POMC mRNA using the Amersham DNA Colour Kit and the OmniSlide System. (c) Detection of POMC mRNA using the Amersham RNA Colour Kit and conventional instrumentation. (d) Detection of POMC mRNA using the Amersham RNA Colour Kit and the OmniSlide System.

3.5.2 Absolute control of sample temperature

The extent to which sample temperature is controlled by the OmniSlide and TouchDown control software can be demonstrated by looking at the denaturation step in DNA ISH protocols. Where DNA is the target sequence it is essential that the denaturation step is performed at its optimal temperature for the specified length of time. Inadequate denaturation resulting from too low an incubation temperature will generate weaker than expected signals. Excessive denaturation, resulting from too high a temperature, will compromise sample morphology. It is also important that the whole area of the sample on the slide is subjected to the same optimal denaturation conditions; lack of uniform heating across the sample will lead to 'patchy' results with varying levels of hybridization signal strength.

The OmniSlide and TouchDown flat blocks have been set up to give accurate control of sample temperature and better than ±1°C uniformity across both the length and width of the blocks. This has been demonstrated using the Cytocell Multiprobe system which allows the simultaneous detection, using FISH, of multiple chromosome abnormalities through the independent hybridization of up to 24 whole-chromosome probes on a single microscope slide (see *Figure 16*). In order to perform the 24 individual reactions, four-fifths of the slide is evenly divided into 24 distinct areas.

The hybridization protocol used with the Chromoprobe Multiprobe system incorporates an optimal 5 min, 75°C denaturation step. Should this temperature drop below 74°C within any area of the slide, the signal intensity achieved will be weaker than expected. As shown in *Figure 16a*, optimal denaturation is achieved across the whole slide giving consistent signal intensity and sample morphology for all 24 sample areas and confirming the uniformity specification of the OmniSlide flat block.

3.5.3 PCR ISH protocols

Hybaid's *In Situ* Systems were developed to cater for every step of a PCR ISH protocol, from the sample permeabilization to the signal detection stages. Since their introduction both the OmniSlide and TouchDown *In Situ* Systems have been used by researchers worldwide to detect both DNA and mRNA molecules (27–35)

PCR ISH was initially seen as a notoriously difficult technique to perform with a tendency to generate unreliable results in terms of quality and accuracy. With a better understanding of the requirements of each practical step and the availability of dedicated instrumentation not only has the reliability of PCR ISH improved but it has also been possible to generate protocols that can be adopted and optimized in different laboratories. While it is important that attention is paid to each stage of a PCR ISH experiment particular care needs to be taken with the amplification and permeabilization steps. In order to generate specific results the primers used to amplify the target sequence must be optimized in solution before PCR ISH is attempted; the defined conditions can then be directly transferred to the slide.

The sample permeabilization step of PCR ISH must be optimized for each new sample type. Inadequate permeabilization inhibits the diffusion of the reagents into the target cell, and so the amplification efficiency of the target sequence is compromised. In contrast, excessive permeabilization increases

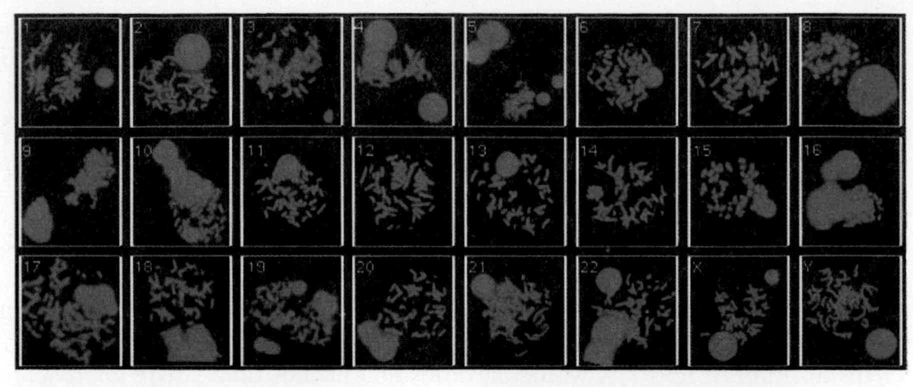

(a)

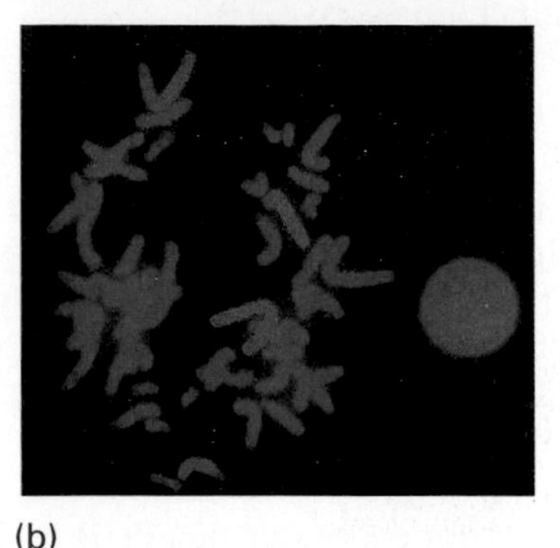

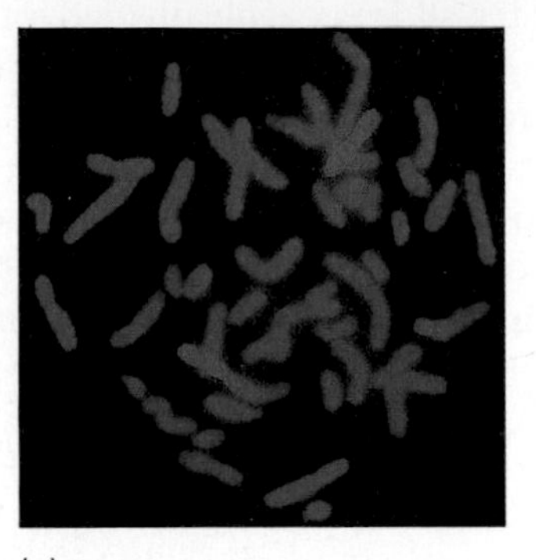

(b) (c)

Figure 16. Detection of chromosomal abnormalities using the Chromoprobe Multiprobe system and the OmniSlide System. A montage (a) is shown containing all 24 chromosome paints from a male patient exhibiting a reciprocal translocation involving chromosomes 1 and 7 using the Cytocell Chromoprobe Multiprobe system; this allows the simultaneous and independent analysis of up to 24 chromosome probes on a single slide. In each case the chromosome is directly labelled with FITC and the cell counterstained with DAPI. All of the chromosome paints appear normal, with the exception of 1 and 7 where the normal and derived chromosomes can be clearly distinguished in (b) and (c) respectively.

the diffusion of the amplified products out of the cell resulting in weak signal intensity and damaged sample morphology. The amplification and permeabilization steps therefore rely upon accurate sample temperature control. This has been demonstrated by Martinez and fellow workers who have developed a robust protocol for performing reverse transcriptase PCR ISH (RT-PCR ISH), optimized on the OmniSlide System, which has been transferred and adopted by different laboratories (31). The specific times and temperatures used for the permeabilization and thermal cycling steps are detailed in *Protocol 1*. An example of where these conditions have been used is demonstrated by the

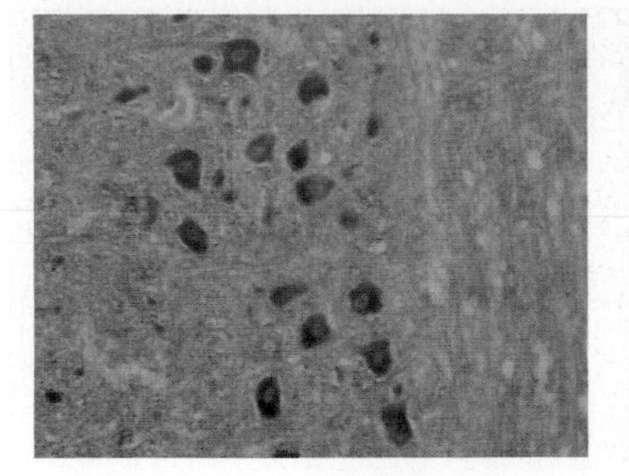

(a)

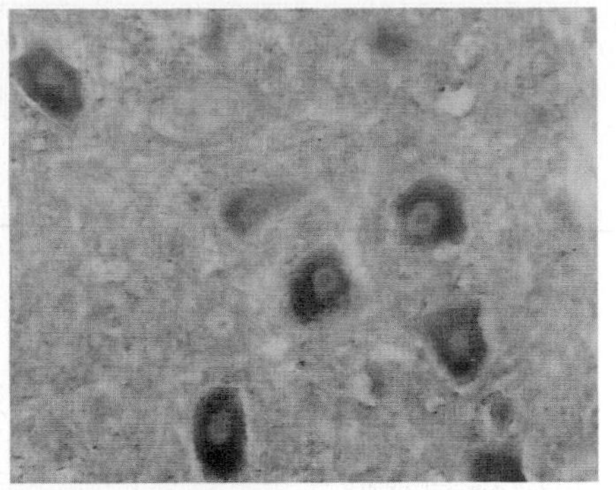

(b)

(c)

Figure 17. Detection of adrenomedullin mRNA in rat brain tissue sections using RT-PCR ISH. Adrenomedullin mRNA was amplified using the rTth enzyme, which has both reverse transcriptase and DNA polymerase functions, and the OmniSlide System. During amplification digoxigenin-dUTP was incorporated into the amplificants which were subsequently detected via the alkaline phosphatase NBT-BCIP colour detection pathway. (a) Test sample: (i) neurones positive for adrenomedullin can be seen with a densely stained, brown/black cytoplasm and a clear nucleus (×100 magnification); (ii) at a higher magnification both positive and negative neurones are visible within the area, showing that adrenomedullin is not expressed in all neurones (×200 magnification). (b) Control sample. Omission of amplification primers from the reverse transcriptase RT-PCR ISH reaction will demonstrate whether the amplification reaction is specific. As can be seen no signal was detected in the neurones on omission of the primers (×200 magnification).

identification of adrenomedullin mRNA in rat brain tissue sections (see *Figure 17*). As demonstrated in *Figure 17* and documented by this group (30–32), the OmniSlide System generates reproducible, specific results with low background levels and good morphology when used to perform PCR ISH protocols.

Protocol 1. Example permeabilization and amplification steps used in RT-PCR ISH

Equipment and reagents

- Proteinase K
- Tris buffer: 0.1 M Tris, 50 mM EDTA, pH 8
- PBS containing 0.1 M glycine
- 5× buffer
- rTth enzyme
- $Mn(OAc)_2$
- dATP, dCTP, dGTP, and dTTP (10 mM each)
- Digoxigenin-dUTP
- Primers

Method

A. *Proteinase K digestion*

1. Incubate sections in a humid chamber with proteinase K in Tris buffer. The concentration of the proteinase K and the incubation time have to be titrated for each tissue. A suitable starting point would be 10 μg/ml and 15 min at 37 °C.
2. Stop the enzyme action by immersing the slides in 0.1 M glycine/PBS for 5 min.
3. Wash in three changes of PBS for 5 min each.

B. *Amplification protocol (combined RT and amplification steps)*

1. Make up RT-PCR ISH solution as follows (75 μl will be needed for each slide; make provision for the control slides too):
   - water 38 μl
   - 5× buffer 15 μl
   - rTth enzyme 3 μl
   - $Mn(OAc)_2$ 7.5 μl
   - dNTPs 2.25 μl each of dATP, dCTP, dGTP, and dTTP
   - digoxigenin-dUTP 1 μl
   - primers 0.75 μl of each
2. Apply 75 μl of RT-PCR ISH solution to each slide.
3. Apply a SureSeal frame and coverslip.
4. Transfer to the OmniSlide Thermal Cycler and run the following program:
   (a) 60 °C for 30 min (one cycle)
   (b) 94 °C for 2 min (one cycle)
   (c) 94 °C for 45 sec; 55 °C for 15 sec; 60 °C for 60 sec (20 cycles)
   (d) 60 °C for 5 min (one cycle).

For further details see ref. 32

3.6 Summary

In developing the OmniSlide and TouchDown *In Situ* Systems Hybaid acknowledged the need for dedicated *in situ* instrumentation that would allow researchers to perform the amplification step of PCR ISH and to automate each stage of this and other *in situ* protocols. This 'whole systems' approach has improved the reproducibility and quality of results generated, allowed the efficient transfer of established protocols between laboratories and made *in situ* protocols far less laborious to perform. Taken together these factors have encouraged an increasing number of researchers to establish *in situ* techniques within their own laboratories.

3.7 Acknowledgements

Hybaid would like to thank Summer Brunning (Amersham International), Martin Lawrie (Cytocell Ltd), and Alfredo Martinez for supplying the results shown in this section of this chapter.

References

1. O'Leary, J.J., Browne, G., Johnson, M.I., Landers, R.J. *et al.* (1994). *J. Clin. Pathol.*, **47**, 933.
2. Saiki, R.K., Scharf, S., Faloona, F., Mullis, K.B., Horn, G.T., Erlich, H.A., and Arnheim, N. (1985). *Science*, **230**, 1350.
3. Saiki, R.K., Gelfand, D.H., Stoffel, S., Scharf, S.J., Higuchi, R., Horn, G.T., Mullis, K.B., and Erlich, H.A. (1988). *Science*, **239**, 487.
4. Haase, A.T., Retzel, E.F., and Staskus, K.A. (1990). *Proc. Natl Acad. Sci. USA*, **87**, 4971.
5. Nuovo, G.J., MacConnell, P., Forde, A., and Delvenne, P. (1991). *Am. J. Pathol.*, **139**, 847.
6. Nuovo, G.J., Gallery, F., MacConnell, P., Becker, P., and Bloch, W. (1991). *Am. J. Pathol.*, **139**, 1239.
7. Komminoth, P., Long, A.A., Ray, R., and Wolfe, H.J. (1992) *Diagnostic Mol. Pathol.*, **1**, 85.
8. Bagasra, O., Hauptman, S.P., Lischer, H.W., Sachs, M., and Pomerantz, R.J. (1992). *New Engl. J. Med.*, **326**, 1385.
9. Gosden, G. and Hanratty, D. (1993). *Biotechniques*, **15**, 78.
10. Patterson, B.K., Till, M., Otto, P., Goolsby, C., Furtado, M.R., McBride, L.J., and Wolinsky, S.M. (1993). *Science*, **260**, 976.
11. Boshoff, C., Schulz, T.F., Kennedy, M.M., Graham, A.K., Fischer, C., Thomas, A., McGee, J.O'D., Weiss, R.A., and O'Leary, J.J. (1995). *Nature Medicine,* **1**, 1274.
12. Staskus, K.A., Couch, L., Bitterman, P., Retzel, E.F., Zupancic, M., List, J., and Haase, A.T. (1991). *Microbiol. Pathogenesis*, **11**, 67.
13. Atwood, J.G. (1994). Instrumentation for *in situ* PCR. In: *PCR In Situ Hybridisation. Protocols and Applications* (ed. G.J. Nuovo), pp. 403–416. Raven Press, New York.

14. O'Leary, J.J., Chetty, K., Graham, A.K., and McGee, J.O'D. (1996) *J. Pathol.*, **178**, 11.
15. Gall, J.G. and Pardue, M.L. (1969). *Proc. Natl Acad. Sci USA*, **63**, 378.
16. John, H.A., Birnstiel, M.L., and Jones, K.W. (1969). *Nature*, **223**, 582.
17. Komminoth, P. and Long, A.A. (1993). *Virch. Archiv. B (Cell Pathol.)*, **64**, 67.
18. Bagasra, O., Seshamma, T., and Pomerantz, R.J. (1993). *J. Immunol. Methods*, **158**, 131.
19. Zehbe, I., Hacker, G.W., Rylabder, E., Sallstrom, J., and Wilander, E. (1992). *Anticancer Res.*, **12**, 2165.
20. Murphy, D.S., Hoare, S.F., Going, J.J., Mallon, E.E.A., George, D., Kaye, S.B., Brown, R., Black, D.M., and Keith, W.N. (1995). *J. Natl Cancer Inst.*, **87**, 1694.
21. Nicholson, F., Ajetunmobi, J.F., Li, M., Shackleton, E.A., Starkey, W.G., Illavia, S.J., Muir, P., and Banatvala, J.E. (1995). *Br. Heart J.*, **74**, 522.
22. Gosden, J. (1995). *Am. Biotech. Lab.*, April.
23. Gosden, J. and Lawson, D. (1994). *Human Mol. Genet.*, **3**, 931.
24. Gosden, J. and Lawson, D. (1995). *Cytogenet. Cell Genet.*, **68**, 57.
25. Speel, E.J.M., Lawson, D., Hopman, A.H.N., and Gosden, J. (1995). *Human Genet.*, **95**, 29.
26. Speel, E.J.M., Lawson, D., Ramaekers, F.C.S., Gosden, J.R., and Hopman, A.H.N. (1996). *Biotechniques*, **20**, 226.
27. Bettinger, D. and Mougin, C. (1994). *J. Virol. Methods*, **49**, 59.
28. Lin, C.-T., Dee, A.N., Chen, W., and Chan, W.-Y. (1994). *Lab. Invest.*, **71**, 731.
29. Palluy, O., Bendani, M., Vallat, J.-M., and Rigaud, M. (1994). *C.R. Acad. Sci. Paris, Sciences de la Vie/Life Sciences*, **317**.
30. Martinez, A. and Ebina, M. (1995). *The NIH Catalyst*, May/June, pp. 12–13.
31. Martinez, A., Miller, M.-J., Quinn, K., Unsworth, E.J., Ebino, M., and Cuttitta, F. (1995). *J. Histochem. Cytochem.*, **43**, 739.
32. Martinez, A., Miller, M.-J., Unsworth, E.J., Siegfried, J.M., and Cuttitta, F. (1995). *Endocrinology*, **136**, 4099.
33. Teo, I.A. and Shaunak, S. (1995). *Histochem. J.*, **27**, 660.
34. Mee, A.P., Davenport, L.K., Hoyland, J.A., Davies, M., and Mawer, E.B. (1996). *J. Mol. Endocrinol.*, **16**, 183.
35. Mee, A.P., Hoyland, J.A., Braidman, I.P., Freemont, A.J., Davies, M. and Mawer, E.B. (1996). *Bone*, **18**, 295.

A1

List of suppliers

Website and e-mail addresses have been included where these could be identified. Useful sources of contact information via the Worldwide Web are: the *Nature* buyers' guide at http://guide.nature.com/; the laboratory products association at http://www.lpassn.org/; and the biosupplynet biotechnology research products source at http://www.biosupplynet.com/

Advanced Biotechnologies Ltd, 7 Mole Business Park, Randalls Road, Leatherhead, Surrey KT22 7BA, UK. Tel.: +44-1372-360123; Fax: +44-1372-363-263; e-mail: adbio@adbio.demon.co.uk; website: `http://www.adbio.co.uk/`

Amersham

Amersham International plc, Amersham Place, Little Chalfont, Bucks HP7 9NA, UK. Tel.: +44 1 494 544000; Fax: +44-1494-542266. Freephone numbers: Orders: 0800-515313, Fax: 0800-616927; Technical enquiries: 800-616928; website: `http://www.amersham.co.uk/`

Amersham North America, 2636 South Clearbrook Drive, Arlington Heights, IL 60005, USA. Customer service: Molecular science +1-800-323-9750, Fax +1-800-228-8735; Technical service: Tel.: +1-800-341-7543, Fax: +1-847-593-0067, e-mail: alssah@ix.netcom.com; website: `http://www.amersham.com/`

Anderman

Anderman and Co. Ltd., 145 London Road, Kingston-Upon-Thames, Surrey KT17 7NH, UK.

Applied Biosystems, Kelvin Close, Birchwood Science Park North, Warrington, Cheshire WA3 7PB, UK. Tel.: +44-1925-825650, Fax: +44-1925-282502850 (see also Perkin–Elmer)

Barnstead/Thermolyne, 2555 Kerper Blvd., PO Box 797, Dubuque, IA 52001, USA.Tel.: +1-319-556-2241; Toll-free: +1-800-553-0039; Fax: +1-319-556-0695.

Beckman Instruments

Beckman Instruments UK Ltd., Oakley Court, Kingsmead Business Park, London Road, High Wycombe, Bucks HP11 1J4, UK.

Beckman Instruments Inc., PO Box 3100, 2500 Harbor Boulevard, Fullerton, CA 92634, USA.

Becton Dickinson

Becton Dickinson Labware, Two Oak Park, Bedford, MA 01730, USA. Toll-free: +1-800-343-2035; website: http://www.cbpi.com/

Becton Dickinson and Co., Between Towns Road, Cowley, Oxford OX4 3LY, UK.

BioGenex Laboratories, 4600 Norris Canyon Road, San Ramon, CA 94583, USA. Toll-free: +1-800-421-4149 or +1-800-DNA-PURE; e-mail: sales@biogenex.com; website: http://www.biogenex.com

Bio-Rad Laboratories

Bio-Rad Laboratories Ltd., Bio-Rad House, Maylands Avenue, Hemel Hempstead HP2 7TD, UK.

Bio-Rad Laboratories, Division Headquarters, 3300 Regatta Boulevard, Richmond, CA 94804, USA.

Boehringer Mannheim

Boehringer Mannheim UK, Bell Lane, Lewes, East Sussex BN7 1LG, UK. Tel.: 0800-521578; Fax: 800-181087; e-mail: biochem_uk@bmg.boehringer-mannheim.com; website: http://biochem.boehringer-mannheim.com/

Boehringer Mannheim Corporation, Biochemical Products, 9115 Hague Road, PO Box 504, Indianapolis, IN 46250-0414, USA. Tel.: +1-800-262-1640; Fax: +1-800-428-2883.

British Drug Houses (BDH) Ltd, Poole, Dorset, UK.

Cel-Line Associates, Inc., 33 Gorgo Lane, PO Box 648, Newfield, NJ 8344, USA. Tel.: +1-609-697-4590; Toll-free: +1-800-662-0973; Fax: +1-609-697-9728.

Dako

Dako, 6392 Via Real, Carpinteria, CA 93013, USA. Tel.: +1-805-566-6655; Fax: +1-805-566-6688; website: http://www.dakousa.com/

Dako, 16 Manor Courtyard, Hughenden Avenue, High Wycombe, Bucks HP13 5RE, UK. Tel.: 0800-614387; e-mail: dakofacts@dakoukltd.co.uk; website: http://www.dakoltd.co.uk/

Difco Laboratories

Difco Laboratories Ltd., P.O. Box 14B, Central Avenue, West Molesey, Surrey KT8 2SE, UK.

Difco Laboratories, P.O. Box 331058, Detroit, MI 48232–7058, USA.

Du Pont

Dupont (UK) Ltd., Industrial Products Division, Wedgwood Way, Stevenage, Herts, SG1 4Q, UK.

Du Pont Co. (Biotechnology Systems Division), P.O. Box 80024, Wilmington, DE 19880–002, USA.

Enzo Diagnostics Inc., 60 Executive Boulevard, Farmingdale, NY 11735, USA. Tel.: +1-516-694-7070; Toll-free: +1-800-221-7705; Fax: +1-516-694-7501; e-mail: custserv@enzobio.com

Erie Scientific Company, 48 Congress Street, Portsmouth, NH 03801, USA. Tel.: +1-603-431-8410; Toll-free: +1-800-258-0834; Fax: +1-603-431-0860.

European Collection of Animal Cell Culture, Division of Biologics, PHLS Centre for Applied Microbiology and Research, Porton Down, Salisbury, Wilts SP4 0JG, UK.

Fisher Scientific Company, 2000 Park Lane, Park Lane Office Center, Pittsburgh, PA 15275, USA. Tel.: +1-412-490-8300; Fax: +1-412-490-8900 (sales).

Flow Laboratories, Woodcock Hill, Harefield Road, Rickmansworth, Herts. WD3 1PQ, UK.

Fluka

Fluka-Chemie AG, CH-9470, Buchs, Switzerland.

Fluka Chemicals Ltd., The Old Brickyard, New Road, Gillingham, Dorset SP8 4JL, UK.

Gibco BRL

Gibco BRL (Life Technologies Ltd.), Trident House, Renfrew Road, Paisley PA3 4EF, UK.

Gibco BRL (Life Technologies Inc.), 3175 Staler Road, Grand Island, NY 14072–0068, USA.

Hendley, Oakwood Hill Industrial Estate, Loughton, Essex, UK. Tel.: +44-181-502-1821; Fax: +44-181-502-0430.

Arnold R. Horwell, 73 Maygrove Road, West Hampstead, London NW6 2BP, UK.

HT Biotechnology Ltd, Unit 4, 61 Ditton Walk, Cambridge CB5 8QD, UK.

Hybaid, Life Sciences International (UK) Ltd, Unit 5, The Ridgeway Centre, Edison Road, Basingstoke RG21 6YH, UK. Tel.: +44-1256-817282; Fax: +44-1256-817292; e-mail: sharon.osborne@lifesciences.com; website: `http://www.hybaid.co.uk/`

HyClone Laboratories 1725 South HyClone Road, Logan, UT 84321, USA.

International Biotechnologies Inc., 25 Science Park, New Haven, Connecticut 06535, USA.

Invitrogen Corporation

Invitrogen Corporation 3985 B Sorrenton Valley Building, San Diego, CA. 92121, USA.

Invitrogen Corporation c/o British Biotechnology Products Ltd., 4–10 The Quadrant, Barton Lane, Abingdon, OX14 3YS, UK.

Kodak: Eastman Fine Chemicals 343 State Street, Rochester, NY, USA.

Life Technologies

Life Technologies, Inc., 8717 Grovemont Circle, Gaithersburg, MA 20898, USA. Tel.: +1-301-840-8000; Fax: +1-301-670-1394; website: `http://www.lifetech.com/`

Life Technologies, PO Box 35, Trident House, Renfrew Road, Paisley PA3 4EF, UK.

Merck

Merck Industries Inc., PO Box 4, Bldg WP 53, West Point, PA 19486, USA. Tel.: +1-908-594-4600; Toll-free: +1-800-672-6372; Fax: +1-215-652-2953; website: `http://www.merck.com/`

Merck, Merck House, Poole, Dorset BH15 1TD, UK.

Millipore

Millipore (UK) Ltd., The Boulevard, Blackmoor Lane, Watford, Herts WD1 8YW, UK.

Millipore Corp./Biosearch, P.O. Box 255, 80 Ashby Road, Bedford, MA 01730, USA.

MJ Research Inc., 149 Grove Street, Watertown, MA 02172, USA. Tel.: +1-617-923-8000; Toll-free: +1-800-729-2165; Fax: +1-617-923-8080.

New England Biolabs (NBL)

New England Biolabs (NBL), 32 Tozer Road, Beverley, MA 01915–5510, USA.

New England Biolabs (NBL), c/o CP Labs Ltd., P.O. Box 22, Bishops Stortford, Herts CM23 3DH, UK.

Nikon Corporation, Fuji Building, 2–3 Marunouchi 3-chome, Chiyoda-ku, Tokyo, Japan.

Nona Casina, 24 Claremont Place, Newcastle upon Tyne, NE2 4AA. Tel.: +44-141-2228550; Fax.: +44-191-2228687.

Ortho Diagnostic, Inc., 1001 US Highway 202, PO Box 350, Raritan, NJ 08869, USA. Toll-free: +1-800-322-6374.

Perkin–Elmer

Perkin–Elmer Ltd, Lincoln Centre Dr., Foster City, CA 94404, USA. Tel.: +1-415-570-6667; Toll-free: +1-800-345-5224; Fax: +1-415-572-2743.

Perkin–Elmer Ltd, Beaconsfield, UK. Tel.: +44-1494-676161; Fax: +44-1494-679331/9333; e-mail: info@perkin-elmer.com; website: `http://www.perkin-elmer.com/`

Pharmacia

Pharmacia Biotech, 800 Centennial Avenue, Piscataway, NJ 08855-132, USA. Tel.: +1-908-457-8000; Toll-free: +1-800-526-3593, Fax: +1-908-457-0557.

Pharmacia, 23 Grosvenor Road, St Albans, Herts AL1 3AW, UK. Website: `http://www.biotech.pharmacia.se/`

Promega

Promega Corporation, 2800 Woods Hollow Road, Madison, WI 53711-5399, USA. Tel.: +1-608-274-4430; Toll-free: +1-800-356-9526; Fax: +1-608-277-2601.

Promega Ltd, Delta House, Chilworth Research Centre, Southampton, Hants SO16 7NS, UK. Tel.: +44-1703-760225; Freephone: 0800-378994, Fax: +44-1703-767014 or 0800-181037, e-mail: custserv@promega.com website: `http://www.promega.com/`

Qiagen

Qiagen Inc., c/o Hybaid, 111–113 Waldegrave Road, Teddington, Middlesex, TW11 8LL, UK.

Qiagen Inc., 9259 Eton Avenue, Chatsworth, CA 91311, USA.

Research Genetics, 2130 South Memorial Parkway, Huntsville AL 35802, USA. Tel.: +1-205-533-4363; Toll-free: +1-800-533-4363; Fax: +1-205-536-9016; website: `http://www.resgen.com/`

Richard-Allen Medical, 8850 M-89, Richland, MI 616-629-5811, USA.

Sanford Corporation

Sanford Corporation, c/o Berol Limited, Olmedow Road, King's Lynn, Norfolk, PE30 4JR, UK. Tel.: +44-1553-761221; Fax: +44-1553-766534; e-mail: mail@sanford.co.uk; website: `http://www.sandfordcorp.com/` or `http://194.217.46.20/berol/`

Sanford Corporation, 2711 Washington Boulevard, Bellwood, IL 60104, USA. Tel.: +1-708-547-6650; Fax: +1-708-547-6719; e-mail: info@sanfordcorp.com

Schleicher and Schuell

Schleicher and Schuell Inc., Keene, NH 03431A, USA.

Schleicher and Schuell Inc., D-3354 Dassel, Germany. Schleicher and Schuell Inc., c/o Andermann and Company Ltd.

Shandon Scientific Ltd., Chadwick Road, Astmoor, Runcorn, Cheshire WA7 1PR, UK.

Sigma Chemical Co.

Sigma Chemical Co., PO Box 14508, St Louis, MO 63178, USA. Tel.: +1-314-771-5750; Toll-free: +1-800-325-3010; Fax: +1-800-325-5052.

Sigma Chemical Co., Fancy Road, Poole, Dorset BH12 4QH, UK. Tel.: +44-1202-733114; Freephone: 0800-373731; Fax: +44-1202-715460; e-mail: custserv@sial.com; website: `http://www.sigma.sial.com/`

Stratagene

Stratagene Ltd, Cambridge Innovation Centre, Cambridge Science Park, Milton Road, Cambridge CB4 4GF UK. Tel.: +44-1223-420955; Freephone: 0800-585370; Fax: +44-1223-420234.

Stratagene Inc., 11011 N. Torrey Pines Rd, La Jolla, CA 92037, USA. Tel.: +1-619-535-5400; Toll-free: +1-800-424-5444; Fax: +1-619-535-0034.

Streck Laboratories Inc., 14306 Industrial Road, Omaha NE 68144, USA. Tel.: +1-402-333-1982; Toll-free: +1-800-843-0912; Fax: +1-402-333-4094 (sales).

Tip-Top, Stahlgruber, DS-8011 Poing, Germany.

United States Biochemical, P.O. Box 22400, Cleveland, OH 44122, USA.

Vector Laboratories

Vector Laboratories, Inc., 30 Ingold Road, Burlingame, CA 94010, USA. Tel.: +1-415-697-3600, Toll-free: +1-800-227-6666; Fax: +1-415-697-0339.

Vector Laboratories, 16 Wulfric Square, Bretton, Peterborough, UK. Tel.: +44-1733-265530; Fax: +44-1733-263048.

VWR Scientific Products, 1310 Goshen Pkwy, West Chester, PA 19380, USA. Tel.: +1-610-431-1700; Toll-free: +1-800-932-5000; Fax: +1-610-936-1761; website: `http://www.vwrsp.com/`

Wellcome Reagents, Langley Court, Beckenham, Kent BR3 3BS, UK.

Index

Index